Build
Your Own
Polyhedra

State University of New York, Binghamton

Jean Pedersen
Santa Clara University

DALE SEYMOUR PUBLICATIONS
Pearson Learning Group

Acknowledgments

We would like to express our deep gratitude to all those, too numerous to mention by name—students, teachers in summer programs, and others—who have helped us greatly by trying out our ideas and reporting on the results. They have contributed significantly to making this a better book.

Design: Mark Ong
Cover Design: Christy Butterfield
Illustrations: Charlotte Carlisle
               Mambi Luis
Photographs: Wayland Lee

ISBN 0-201-49096-X
Printed in the United States of America
    11 12 13   07 06 05

1-800-321-3106
www.pearsonlearning.com

About the Authors

Peter Hilton is Distinguished Professor of Mathematics at the State University of New York at Binghamton. Jean Pedersen is Associate Professor of Mathematics at Santa Clara University, Santa Clara, California.

Professor Hilton is past Chairman of the U.S. Commission on Mathematical Instruction and past Secretary/Treasurer of the International Commission on Mathematical Instruction. He is the author of several books and articles on algebraic topology, homological algebra, and group theory.

Professor Pedersen is past Governor of the Northern California Section of the Mathematical Association of America. She received an award from SCU for "Outstanding achievement in teaching, research, and service in the Department of Mathematics." She is the author of several articles and books on geometry.

Together Professor Hilton and Professor Pedersen have written several articles on mathematical research and contributions to the study of problems of mathematics education. They are joint authors of *Fear No More: An Adult Approach to Mathematics* and, with Jean Benson, *College Preparatory Mathematics*.

Dedicated, in deep affection,

to the memory of

Peter's mother, Betty Hilton

and

Jean's father, Ralph E. Jorgenson

Contents

I hear, and I forget,

I see, and I remember,

I do, and I understand.

(Chinese Proverb)

Introduction

This book is different! We don't suggest that you should read it straight through from the beginning to the end. It is true that we have organized the topics in a logical order, with the ideas that belong together being presented together, so we don't recommend that you merely dip into various chapters at random. But human beings do not usually acquire information in an orderly linear fashion. We tend to learn something, play with it for a while, and use our new information to do things we have not been able to do before. Later we return to the original idea and study it more carefully. Then, when we don't see what more we can do with that particular idea, we move on. In a nutshell, this is what you will probably do with this book.

Chapter 1 is crucial. If you know how to fold all the polygons described in Chapter 1, then the constructions described in Chapters 3 through 11 may be done in any order. However, we believe that there is a natural order, both in terms of a person's expected learning pattern and in terms of his or her visual imagination and development of manipulative skill. These considerations are reflected in the arrangement of chapters (thus 2-dimensional constructions precede 3-dimensional constructions) and also in the arrangement of topics within a chapter.

The order of events in Chapters 3 through 11 shows that we have presented the easiest constructions first. Thus, although it is true that the collapsoids of Chapter 11 require only equilateral triangles in their construction, we have left them until late in the book because we believe they will be easier to understand at that point. Nevertheless, a highly motivated person could certainly construct them—without, perhaps, fully appreciating their properties—as soon as he or she has learned how to fold equilateral triangles, as described in Sections 1.2 and 1.3.

We suggest that you start with Chapter 1 and continue folding the various types of polygons, taking excursions into other chapters to build models as the urge strikes you. We would especially urge you to look seriously at Chapter 2 at some stage of your reading to learn how to fold other polygons that are not explicitly discussed in Chapter 1 and to see the richness of the mathematics connected with folding polygons. Likewise, we encourage you, once you have made your models, to look carefully at Chapter 12. What you will discover is that your beautiful models will become even more beautiful when you understand the mathematics connected with them.

Who Are Our Readers?

The obvious logical answer is, You are! However, let us elaborate a little on our intended readership. We have chosen an expository style that should, we believe, make this book accessible—and attractive—to any intelligent person aged between twelve and one hundred. It does not need the intervention of a teacher to mediate between the text and the student; but we would hope that many teachers will read, understand, and appreciate our text and then pass on their understanding and their enthusiasm to their students. Thus we particularly hope that senior high, junior high, and even upper-grade elementary school teachers will read this book and incorporate the appropriate parts of its content into their own teaching. Naturally, then, we also hope that our book will be appreciated by those concerned with both in-service and pre-service education of teachers and those responsible for summer institutes for teachers. We need the support and active interest of all these professionals to restore the study of geometry to the place it should have in the minds and hearts of all sensitive, intelligent people.

Should You Always Follow Instructions?

Any self-respecting human being, and therefore all our readers, must answer this question with a resounding NO! In the next paragraph we describe two aspects of our building instructions where we do advise rather rigid adherence to our specifications. However, we are very far from recommending that you fold all your regular polygons and construct all your polyhedra exactly as described. What we have done is to give you *algorithms* for the relevant constructions. Machines follow algorithms with relentless fervor; human beings look for special ways of doing particular things. Always feel free to use your ingenuity to avoid an algorithm that is not working for you. (This, of course, is exactly the advice teachers should give children learning to do computations!)

A Word to the Wise

We've done a lot of field testing of the material in this book. Our instructions seem to be, on the whole, quite comprehensible to most readers. However, there are two basic types of error that people seem prone to make in following the instructions.

Material Error In doing mathematics, it is absurd to specify the type or quality of paper on which the mathematics should be done. However, when we describe to you how to make mathematical models, we must insist that the choice of material is not arbitrary—instructions for making models that are easily constructed using gummed mailing tape are unlikely to be effective if a strip of paper taken from an exercise book is used! Sometimes it may merely be a question of the finished model not being sufficiently sturdy, but it may even be true that the instructions simply cannot be carried out with inappropriate materials. Exercise your own initiative in choosing which models to make but not in your choice of material (except within very narrow limits).

Geometrical Error Look carefully at the two illustrations at the top of the next page. Do you see a difference? If you do not see a difference, *look again!*

   Notice that in (a) the portion of the strip going in the downward direction is *on top* of the horizontal part of the strip; whereas in (b) that portion is *underneath* the horizontal part of the strip. You will save yourself a great deal of time and effort if you will accustom yourself to looking very carefully at the illustrations, especially with respect to this distinction.

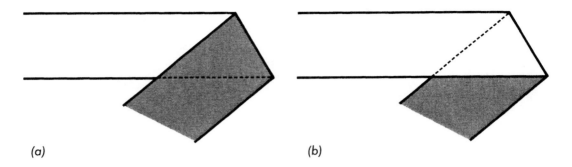

(a)   (b)

Here are some further examples about how to interpret the illustrations.

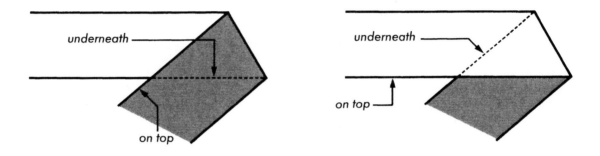

When a strip of paper is folded along a crease line, we indicate the revealed part of the *back* of the paper by shading.

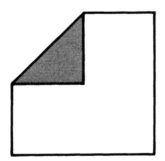

The following symbols instruct you to turn the paper over in the direction of the arrow.

For example, here is how the original and the turned piece would look, in our illustrations, for a transparent piece of plastic with an upper case *F* printed on it; we show two ways of turning the plastic piece.

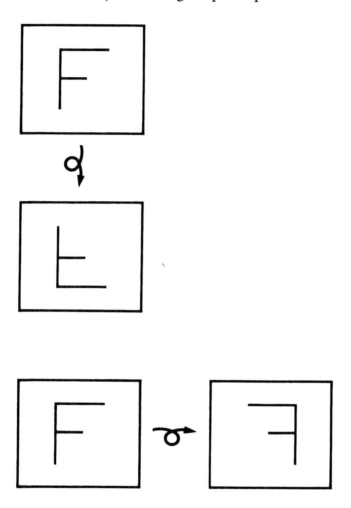

The following arrow means that by performing the indicated moves on the left-hand figure, we obtain the right-hand figure.

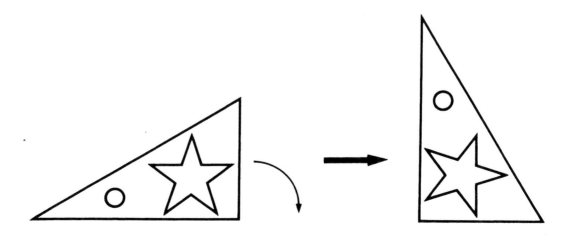

The Expected Effects on Our Readers

Please notice that we are talking of the "expected effects" rather than the "hoped for effects." This is because we are incorrigible optimists—were we not, we would never have undertaken the labor of writing a book in which precision and detail are so vital. Notice, too, that we do not talk of the "effects of reading our book." This, of course, is because you are expected to do much more than merely read the pages that follow. The Chinese proverb

> I hear, and I forget,
> I see, and I remember,
> I do, and I understand.

is true in all learning situations but especially in the two situations covered by this book, namely, when you are learning how to build something and when you are learning mathematics. Even the professional standards of the NCTM encourage teachers to "engage students in an active process of learning in which the students create, discover, and make sense of mathematics."

Certainly we want you to be able to build fascinating geometric models and to understand some of the mathematics that goes with them. This we might call our *local* purpose. By contrast, our *global* purpose is to encourage an appropriate attitude towards mathematics itself. Many children carry into adulthood the false view that it is in the nature of mathematics to proceed from assigned task, via prescribed methods, to the unique "right answer." We expect our readers to enjoy taking the initiative in inventing new problems (often by modifying familiar problems) and new rules of procedure, and to welcome the multiplicity of possible outcomes. We have drafted our text with a view to whetting your appetite for further work, both in constructing geometrical models and in mathematics, arising directly or indirectly from the experiences offered to you in the following pages. Remember that asking good questions is as important as answering them; and experiment is the road to knowledge.

GOOD LUCK!

Using the Metric System

You may convert the measurements in this book to the metric system by using the conversions:

$$1 \text{ in.} = 2.54 \text{ cm} = 25.4 \text{ mm}$$
$$1 \text{ ft} = 30.48 \text{ cm}$$
$$1 \text{ yd} = 0.914 \text{ m}$$

1 *Folding Regular Polygons*

1.1 *Some History*

The Greeks were fascinated with the challenge of constructing regular convex polygons—that is, those polygons in which all sides are of the same length and all angles are equal. They wanted to create these polygons using only an unmarked straightedge and a compass. Drawing an exact geometric figure with these restrictions is called a *Euclidean construction*, and the straightedge and compass are called *Euclidean tools*. The Greeks were successful around 350 B.C. in devising Euclidean constructions for regular convex polygons having the following number of sides:

$$3, 2 \times 3, 4 \times 3, 8 \times 3, \dots \quad \text{(or } 2^n \times 3, \text{ where } n \geqslant 0)$$
$$5, 2 \times 5, 4 \times 5, 8 \times 5, \dots \quad \text{(or } 2^n \times 5, \text{ where } n \geqslant 0)$$
$$15, 2 \times 15, 4 \times 15, \dots \quad \text{(or } 2^n \times 15, \text{ where } n \geqslant 0)$$
$$4, 8, 16, \dots \quad \text{(or } 2^n, \text{ where } n \geqslant 2)$$

No further progress seems to have been made in the next two thousand years, until Gauss (1777–1855) discovered that a Euclidean construction of a regular convex polygon is possible if and only if the number of sides of the polygon is expressible as a power of 2 times a product of distinct Fermat primes.* This discovery implied that, in addition to what the Greeks could do, it is (theoretically!) possible to construct, with Euclidean tools, regular convex polygons having the following number of sides:

$$17, 2 \times 17, 4 \times 17, \dots$$
$$257, 2 \times 257, 4 \times 257, \dots$$
$$65537, 2 \times 65537, 4 \times 65537, \dots$$
$$3 \times 17, 2 \times 3 \times 17, 4 \times 3 \times 17, \dots$$
$$3 \times 257, 2 \times 3 \times 257, 4 \times 3 \times 257, \dots$$
$$3 \times 65537, 2 \times 3 \times 65537, 4 \times 3 \times 65537, \dots$$
$$5 \times 17, 2 \times 5 \times 17, 4 \times 5 \times 17, \dots$$
$$\vdots$$
$$3 \times 5 \times 17 \times 257 \times 65537, 2 \times 3 \times 5 \times 17 \times 257 \times 65537,$$
$$4 \times 3 \times 5 \times 17 \times 257 \times 65537, \dots$$

but no others. (The only known Fermat primes are 3, 5, 17, 257, and 65537.)

---

*A Fermat prime, F_n, is a prime number of the form $F_n = 2^{2^n} + 1$. Thus, $F_0 = 3$, $F_1 = 5$, $F_2 = 17$, $F_3 = 257$, and $F_4 = 65537$ are Fermat primes. Euler (1707–1783) showed that F_5 is *not* a prime number, and that, in fact, $F_5 = 2^{2^5} + 1 = 2^{32} + 1$ is the product 641×6700417. It is not known whether or not any other Fermat numbers F_n are prime. It is known (with the help of a large computer) that F_6 is not prime.

It is unlikely that you will be interested in constructing regular polygons with as many sides as most of those in this list. But it may be interesting to observe that some relatively small numbers are not in this list. Thus, for example, if you had your heart set on constructing a regular convex polygon with 7 or 9 sides, you would know from Gauss' result that this is impossible using only Euclidean tools. Of course, Gauss might well have been able to tell you how to construct such a polygon using more sophisticated tools—and he might have had interesting methods of obtaining very respectable *approximations*. You will be learning much more about this idea of approximate construction in the pages that follow.

Some regular polygons have special names, which are based (except for the first few) on the Greek names for the numbers. We cite just a few examples in the following table.

N = number of sides	Name of regular polygon having N sides
3	equilateral (or equiangular) triangle
4	square (or regular quadrilateral)
5	regular pentagon
6	regular hexagon
7	regular heptagon
8	regular octagon
9	regular nonagon
10	regular decagon
12	regular dodecagon

Notice that the word *equilateral* appears only with the triangle. Do you see why it is not sufficient in the other cases to describe a regular polygon as equilateral (or, for that matter, equiangular)?

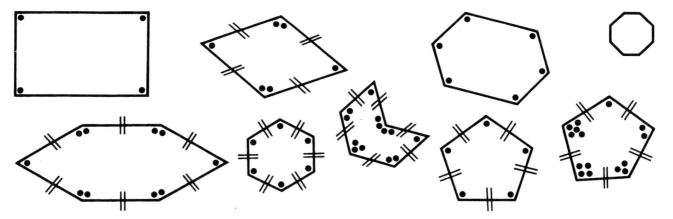

You'll be happy to know that it is not necessary to memorize *all* those names— that's not what geometry, or mathematics, is about! In this book we will only refer to the special names of polygons when they have 3, 4, 5, 6, or 8 sides, and even then it should be clear from the context how many sides are involved. To make the terminology simpler for you we will adopt the convention that

a regular convex polygon with N sides is called a *regular N-gon*.

Thus a regular N-gon is convex unless otherwise stated (there are also regular *star* polygons, which we talk about later, but these are not polygons in the strict sense). Naturally our convention only makes sense when N is a number greater than or equal to 3. (What is a 2-gon?)

The Euclidean constructions of regular polygons are *exact*—that is, they are perfect *in the mind*. However, the *real* accuracy of any actual Euclidean construction on paper is, and owing to the nature of reality must always be, imperfect, depending on many variables. Some of the factors that affect the degree of accuracy in the finished construction are the smoothness of the paper or surface on which the construction is drawn, the sharpness of the drawing instrument (pen, pencil, chalk, stick, and so on), and, of course, the skill of the person making the construction. This last factor affects many young children (and even some adults).

"Don't spoil my circles!"

Archimedes in 212 B.C.

There is, however, a way to construct regular polygons by simply folding a straight strip of paper. In this chapter we show you how this may be done. In fact, the illustrations on the following pages show you very explicitly how to fold regular 3-, 4-, 5-, 6-, 7-, 8-, 9-, and 10-gons. The discussion in the next chapter describes how you may devise folding techniques to obtain certain other regular N-gons. Moreover, we make the fold lines simply by guesswork or by bisecting angles—we never use a protractor to measure angles.

Just as with any kind of construction, the perfection of the finished product depends on your skill. However, the constructions we describe are not nearly as sensitive to your ability to work accurately as Euclidean straightedge and compass constructions. The constructions of the Greeks were theoretically exact but very rough in practice; the constructions we describe are theoretically approximate but very accurate in practice.

Another way in which these paper-folding techniques are different from the Euclidean constructions is that their justification requires practically no knowledge of formal geometry. However, if you do have some knowledge of elementary plane geometry, don't let that distract you during the actual paper-folding. We suggest that you wait until you've completed your folding of the polygon before you try to figure out *why it works.*.

This means that we are asking you to trust us for the moment and follow the instructions carefully. You will be able to construct some very accurate regular polygons with surprisingly little effort even if you don't yet understand why the constructions work.

It is important to begin, so get a strip of paper and follow the step-by-step illustrations for constructing certain regular polygons. We suggest you construct your regular N-gons in the order in which we've arranged the instructions. As you are about to see, in this case, the order 3, 6, 5, 10, 9, 7, 4, 8 is much more natural and enlightening for carrying out these constructions than the more usual ordering of the numbers between 3 and 10.

Along the way we may ask some questions (and in some cases we give some hints as to how those questions may be answered), but you need not worry about these questions unless they interest you. Don't let it worry you if, at first, you want to skip over them. In the next chapter we discuss some of the mathematics connected with regular polygons and with some of our general folding procedures. We include this material for those of you who wish to become more familiar with the polygons you have constructed but, as we've said, you do not need to read or understand all the mathematical parts of this book in order to be able to enjoy constructing the models. Many will prefer to postpone the study of the mathematics to a second reading.

1.2 Preparing to Fold Polygons

Required Materials

☐ Strips (or a roll) of gummed mailing tape, adding machine tape, or brightly colored tape about 1½ in. wide. The glue on the gummed tape should be of the type that needs to be moistened to become sticky. Don't try to use tape that is sticky to the touch when it is dry, unless you want an exercise in frustration.
☐ Transparent tape or white glue, but only if your folding tape is not gummed

Optional Materials

☐ Scissors
☐ Colored acetate cut in 1-in. strips (for teachers only)
☐ Masking tape

A Hint to Teachers

An effective way to teach this activity to your classes is to make strips in bright colors by cutting up acetate report covers. Then, attaching the left-hand end of the colored strip to the glass on the overhead projector with a piece of masking tape, execute the steps outlined in the instructions of this section with that strip. Have your students imitate each of your folds at their own desks with strips of gummed mailing tape.

If you wish to make the rather attractive models shown on the cover of this book, a Polyhedra Kit is available from The Diffraction Company, Inc. (Box 151, Riderwood, Maryland 21139) at a substantial but reasonable price. One kit contains enough materials to decorate two complete sets of the models in *Build Your Own Polyhedra*.

1.3 Folding Triangles and Hexagons

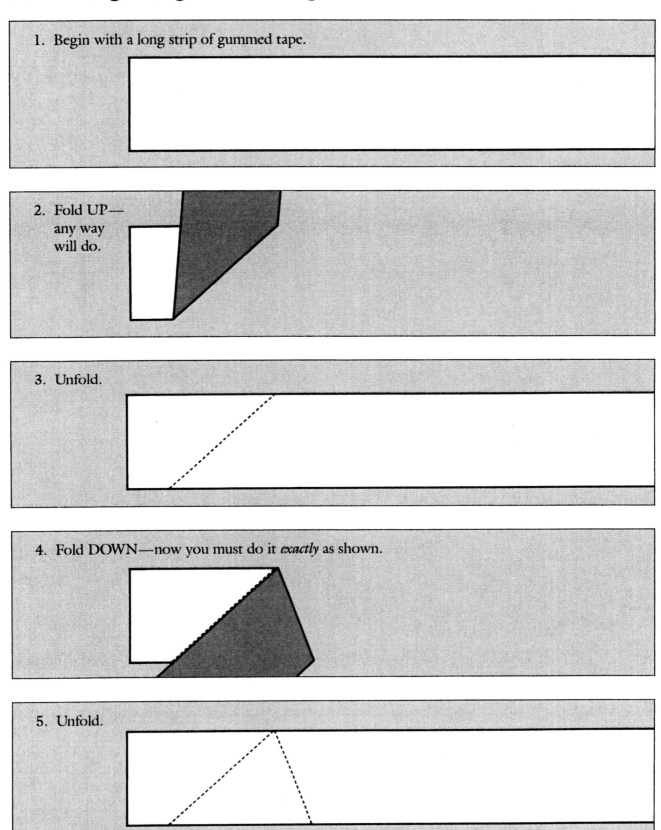

1. Begin with a long strip of gummed tape.

2. Fold UP—
   any way
   will do.

3. Unfold.

4. Fold DOWN—now you must do it *exactly* as shown.

5. Unfold.

6. Fold UP, *exactly* as shown.

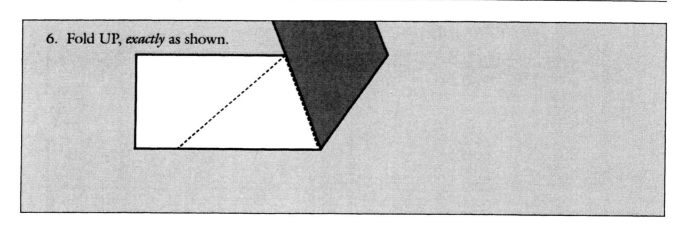

7. Unfold.

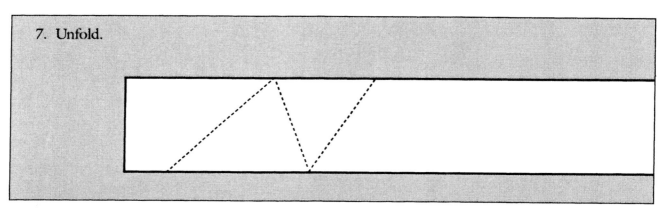

8. Fold DOWN, *exactly* as shown.

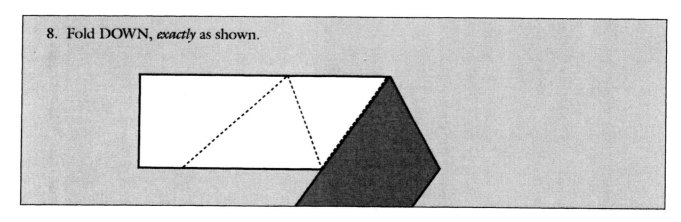

9. Unfold.

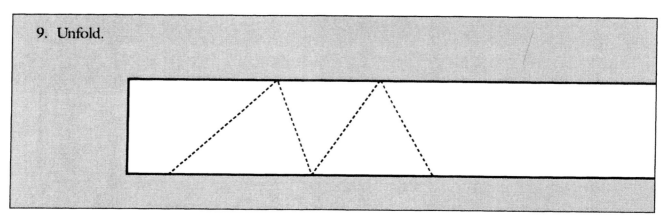

10. Fold UP, *exactly* as shown.

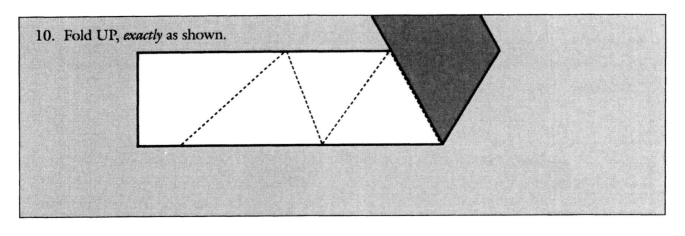

11. Unfold.

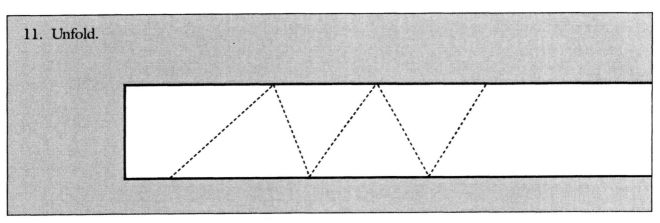

12. Continue folding to make a string of triangles as long as you need. Notice two
things. First, the folding process goes UP, DOWN, UP, DOWN, UP, DOWN...
(which we can abbreviate *UDUDUD*... or U^1D^1). Second, although the first few
triangles may be a bit irregular, the triangles formed always become more and
more regular. (Can you prove it?) When you use these triangles for constructing
models, throw away the irregular ones at the beginning of the tape.

Suppose you want a *bigger* triangle. This can be achieved by taking a strip of
about 30 triangles and executing the F-A-T- (**Fold-And-Twist**) algorithm along the
top of the tape at each of the heavy dots. The F-A-T algorithm is illustrated very
precisely in Figure 1.1. We advise you to master this systematic algorithm, since you
will be using it frequently to construct other polygons.

Of course, you can vary the size of the finished triangle by taking any sequence of
equally spaced dots along the top edge of the tape at the points where the fold lines
meet the top edge.

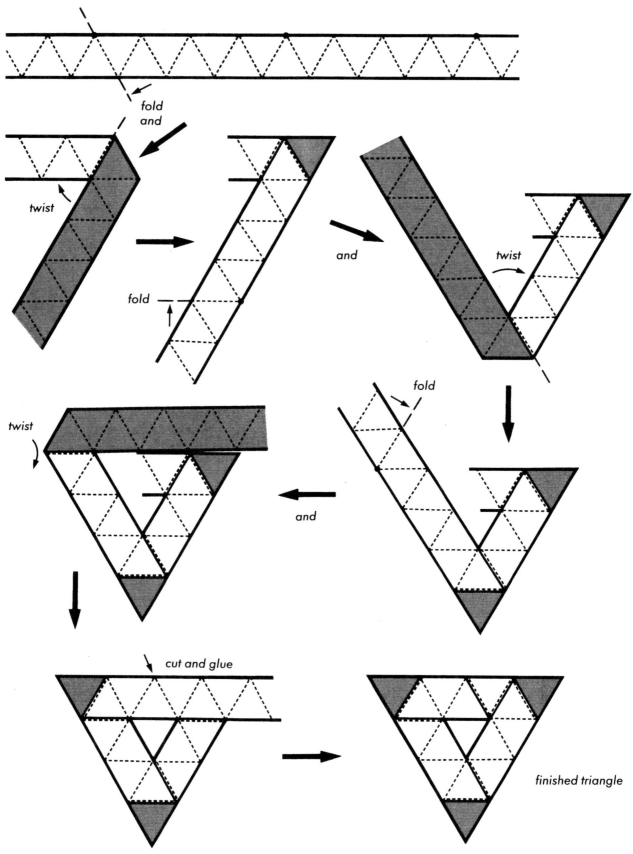

Figure **1.1**   F-A-T algorithm.

A fascinating, versatile toy, called a *flexagon*, can be made from a strip of 10 equilateral triangles. Before folding and gluing your flexagon (as shown in Figure 1.2), be certain to crease all the fold lines in *both* directions (so that the paper flexes easily along each fold line). See Section 3.1 for details about how to make your flexagon work and for instructions about building even bigger and more remarkable flexagons.

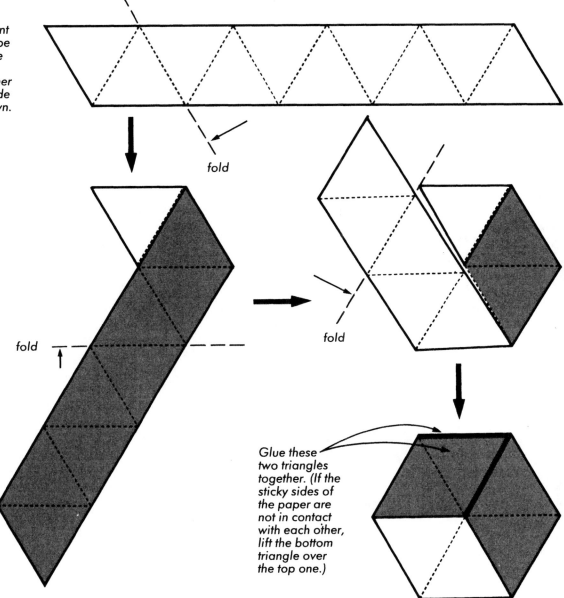

It is important that your tape look just like this. Don't worry whether the sticky side is up or down.

fold

fold

fold

Glue these two triangles together. (If the sticky sides of the paper are not in contact with each other, lift the bottom triangle over the top one.)

Figure **1.2**   Constructing a flexagon.

We have just shown you how to construct a rather special hexagon, using the fact that a regular hexagon may be subdivided into 6 equilateral triangles. Another construction of a regular hexagon may be carried out by adding some *secondary* fold lines to a strip of equilateral triangles (obtained, you will recall, by folding *UDUDUD*...). Try introducing secondary fold lines on a strip of equilateral triangles, as shown next.

1. Begin with a strip of equilateral triangles.

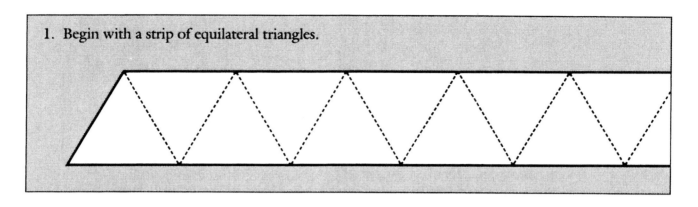

2. Fold down, *exactly* as shown. We call this a *secondary* fold through the top vertex.

3. Repeat step 2 at each of the vertices marked with a heavy dot.

4. Your strip of paper should then look like this:

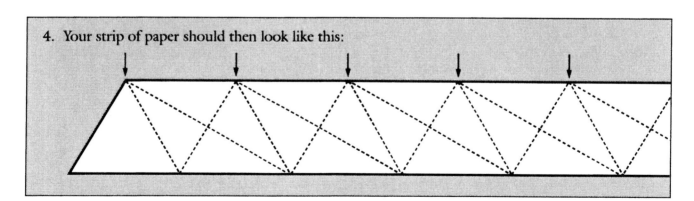

Now do the F-A-T- (fold-and-twist) algorithm at each vertex along the top of the tape indicated by the arrows to obtain a regular 6-gon (with a hole in the center), as shown next.

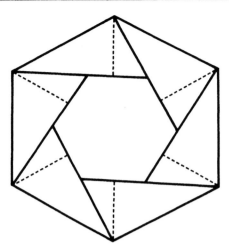

A bigger hexagon can be obtained by increasing the distance between the successive vertices along the top of the tape at which you make your secondary folds. The hexagon is then formed, as before, by performing the F-A-T algorithm at 6 successive vertices equally spaced along the top of the tape. For example, this strip of tape

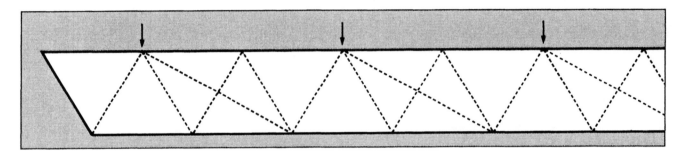

produces the hexagon in Figure 1.3.

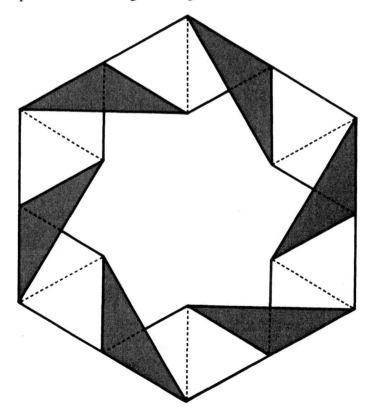

Figure **1.3**

You have seen how to take tape that produces 3-gons (by the F-A-T algorithm) and convert it into tape that produces 6-gons (by the F-A-T algorithm). Can you guess how to add more fold lines so that this tape can be used to produce 12-gons (by the F-A-T algorithm)? Try it!

1.4 Folding Pentagons and 10-gons

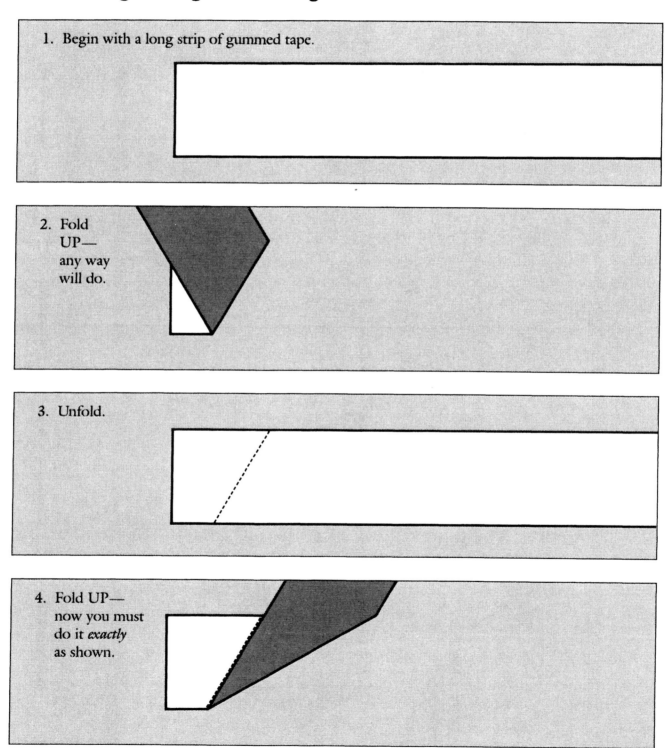

1. Begin with a long strip of gummed tape.

2. Fold UP— any way will do.

3. Unfold.

4. Fold UP— now you must do it *exactly* as shown.

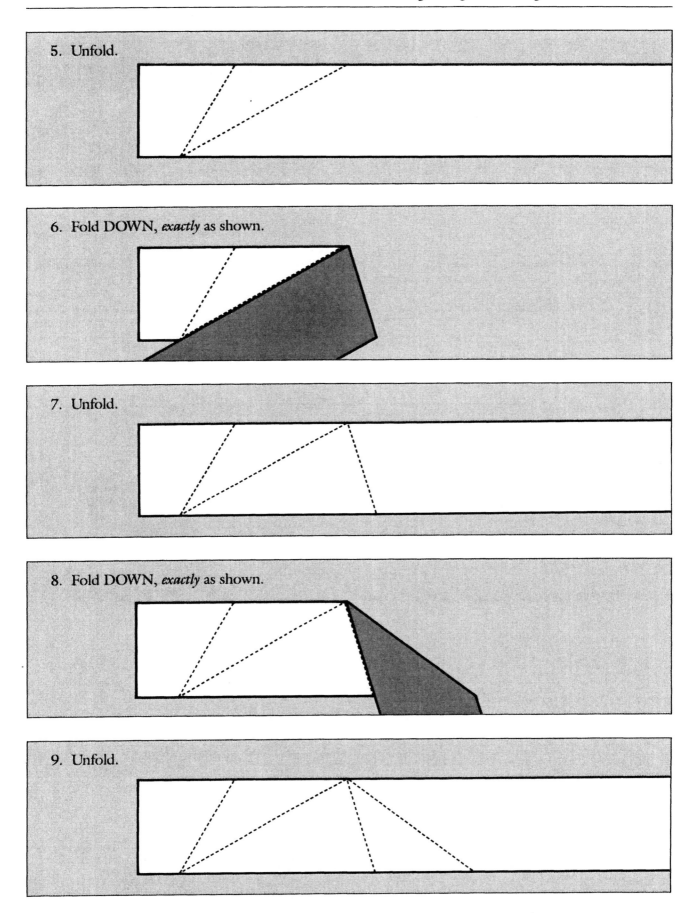

5. Unfold.

6. Fold DOWN, *exactly* as shown.

7. Unfold.

8. Fold DOWN, *exactly* as shown.

9. Unfold.

10. Fold UP, *exactly* as shown.

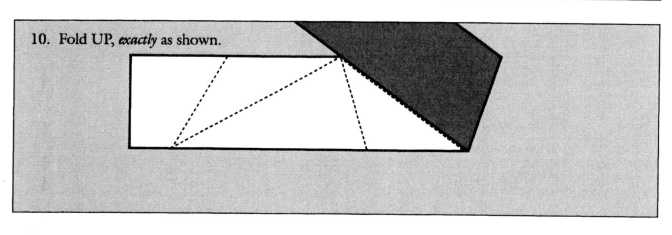

11. Unfold.

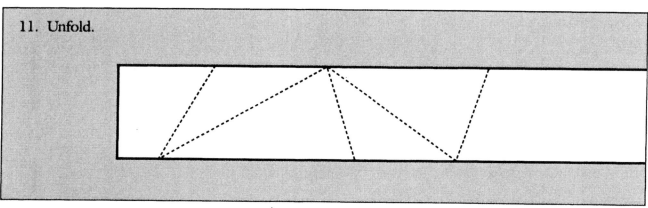

12. Fold UP, *exactly* as shown.

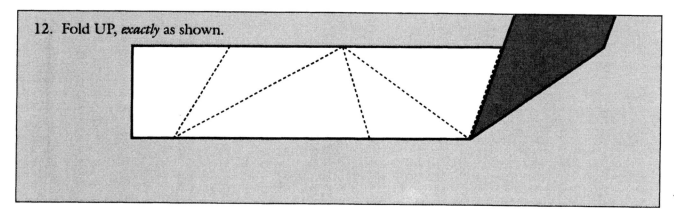

13. Unfold.

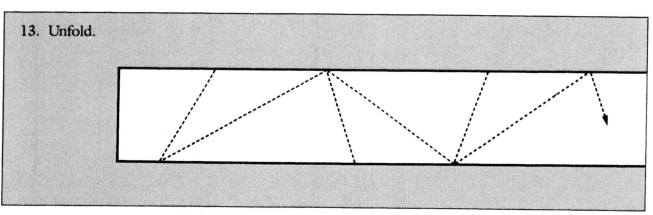

14. Continue folding until you get a few feet of triangles. Notice that the folding process goes UP, UP, DOWN, DOWN, UP, UP, DOWN, DOWN, . . . (which we abbreviate *UUDDUUDD...*, or U^2D^2). Notice, also, that as you fold, the pattern becomes more and more regular. Throw away the first few triangles and try the following construction:

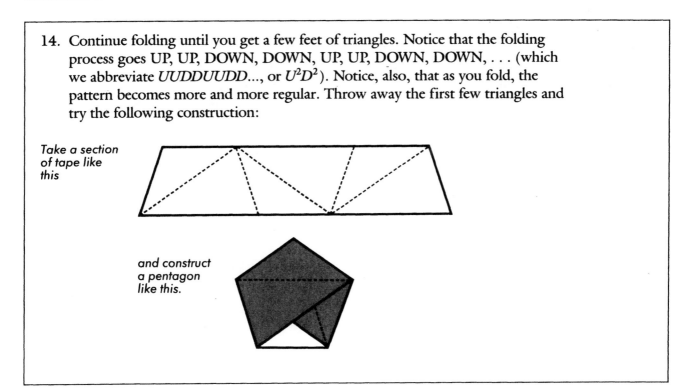

Take a section
of tape like
this

and construct
a pentagon
like this.

This tape, which we describe as having been folded *UUDDUUDD...* (abbreviated U^2D^2), may be used to construct regular pentagons in at least three different ways.

It is important to observe that the tape has *two* kinds of fold lines, some short and some long. If you take this tape and crease it only on the short fold lines (leaving the long fold lines flat), then you obtain the pentagon shown in Step 14. But suppose you crease this tape only on the long fold lines and leave the short fold lines flat. What happens? Try it. If you have difficulty, follow the illustrations in Figure 1.4.

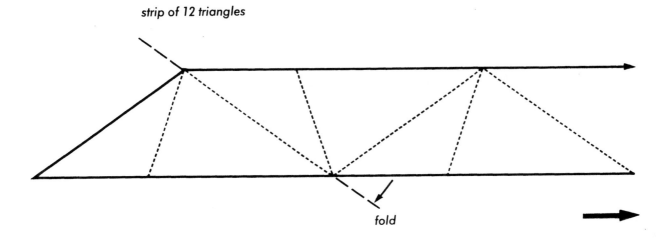

strip of 12 triangles

fold

Figure **1.4**   The *long*-line construction of a pentagon.

Figure **1.4** cont.

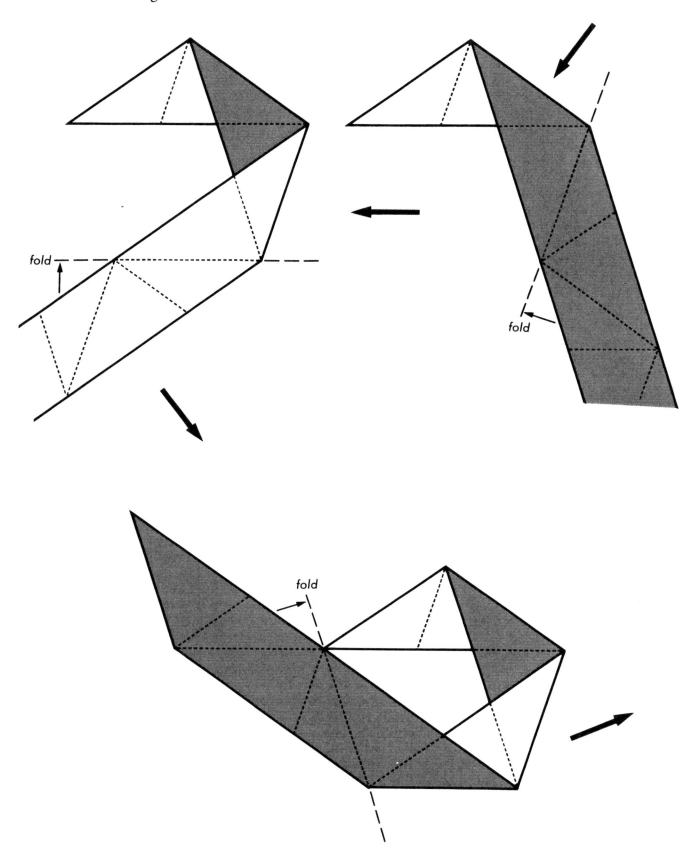

Figure **1.4** cont.

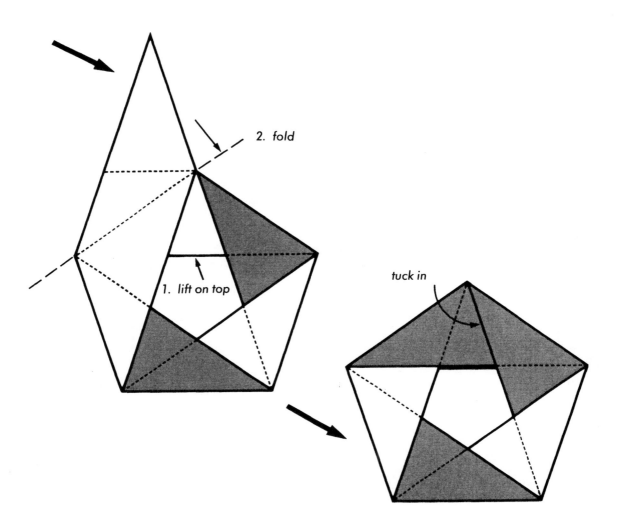

Of course, as you might expect, a regular pentagon can also be folded from this tape using the F-A-T algorithm. The step-by-step illustration of that construction is shown in Figure 1.5.

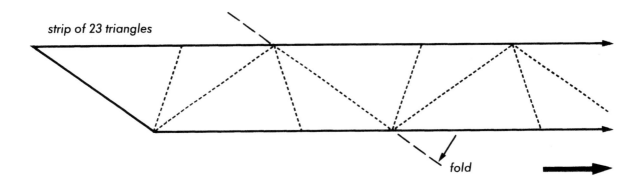

Figure **1.5**   The F-A-T construction of a pentagon.

Figure **1.5** cont.

twist

fold and twist

fold and twist

Figure **1.5** cont.

tuck in

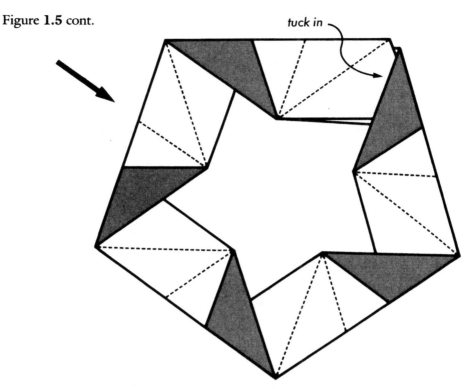

We can also adapt the U^2D^2 tape, by adding secondary fold lines, so that it can be used to construct a regular 10-gon. First, introduce secondary folds as shown next.

1. Begin with a strip of gummed tape folded U^2D^2.

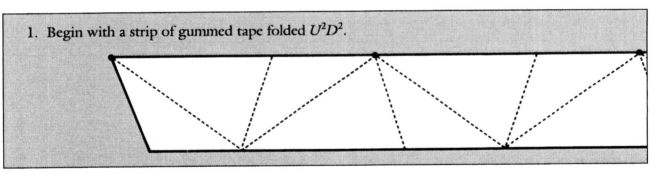

2. Introduce a secondary fold line *exactly* as shown.

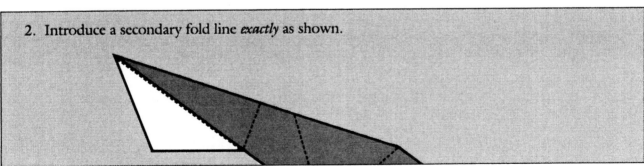

3. Unfold and repeat step 2 at each of 10 equally spaced intervals along the top of the tape. The locations of the desired secondary fold lines are indicated by heavy dots in step 1.

Figure 1.6 shows how the tape should look when you have introduced the necessary secondary folds. Now take this tape and perform the F-A-T algorithm at each of 10 locations along the top edge. The first 4 locations are indicated by heavy dots in Figure 1.6(a). We show only a portion of the 10-gon here, but since you will not be restricted by the size of the pages in this book, you will be able to complete your 10-gon. You will note, however, that when the polygon gets this big it is a little floppy. It may help to place the finished model between two large pieces of newspaper and gently iron it with a steam iron to make it lie flat.

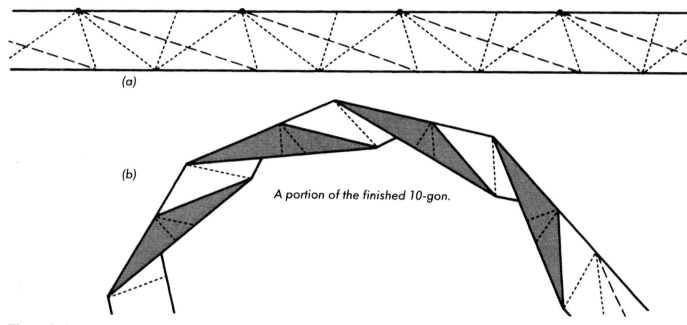

(a)

(b)

A portion of the finished 10-gon.

Figure 1.6

Now, as a reward for following this rather long construction, we will tell you about a particularly easy way to obtain a single pentagon. Just take a strip of paper and begin to tie a knot. As you pull the knot tight, press it flat. See Figure 1.7.

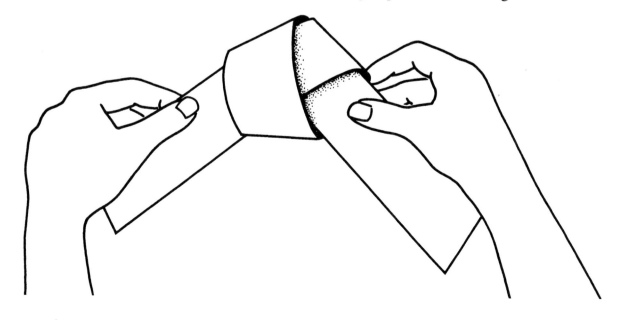

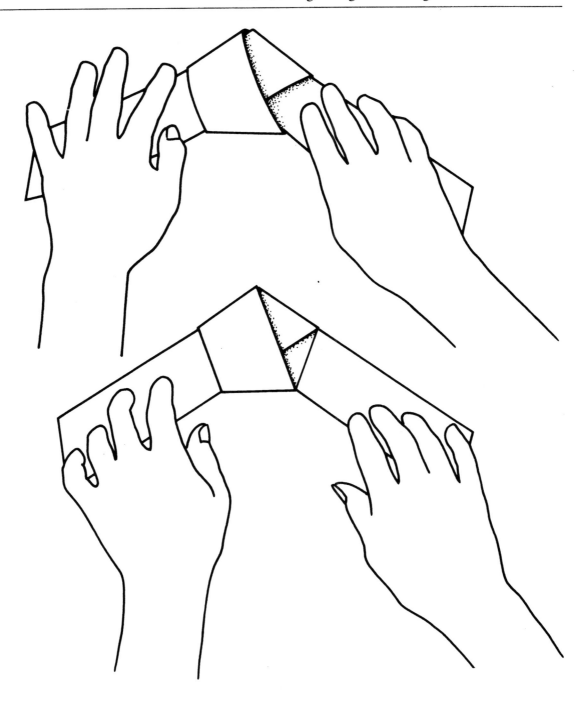

Figure 1.7

In this and the previous section (Section 1.3), we have started to describe a systematic folding procedure, where we make the same number of folds at the top of the tape as at the bottom of the tape. Furthermore, each of the fold lines bisects the angle, on the right, between the last fold line and an edge of the tape. Thus, as you may observe, all new fold lines will "go from left to right," sloping *up* if they are produced by an UP fold and sloping *down* if they are produced by a DOWN fold. If you keep this observation in mind, you can then simply "read off" the folding instructions from any folded strip of tape.

Let us review our results by placing them in a table; see if you can guess a general rule.

By folding tape	and executing the F-A-T algorithm at equally spaced intervals along the top of the folded tape, we obtain a regular polygon having
U^1D^1 U^2D^2 U^3D^3 $\vdots$ U^nD^n	3 sides 5 sides ? sides (make a guess!) $\vdots$? sides (make a guess!)

Let us give you just one bit of information. The correct answer to the first question above is *not* 7 (but that is the most popular *wrong* answer!).

In the next section we show you exactly how to fold tape U^3D^3.... The information we get from that tape will give us the answer to the first question in the preceding table. Of course, after you have discovered and studied the correct answer to the first question, you will be in a much better position to find the correct answer to the second question; you will have the examples $n=1$, $n=2$, $n=3$ from which to try to find the *generalization* for an arbitrary value of n.

1.5   Investigating a Question

1. Begin with a straight strip of gummed tape.

2. Fold UP—any way will do.

3. Unfold.

4. Fold UP, *exactly* as shown.

5. Unfold.

6. Fold UP, *exactly* as shown.

7. Unfold.

8. Fold DOWN, *exactly* as shown.

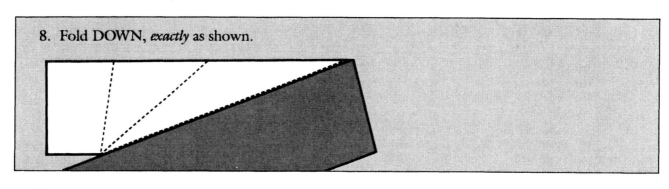

9. Unfold.

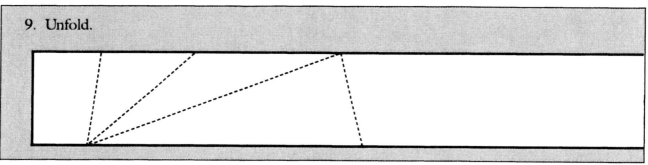

10. Fold DOWN, *exactly* as shown.

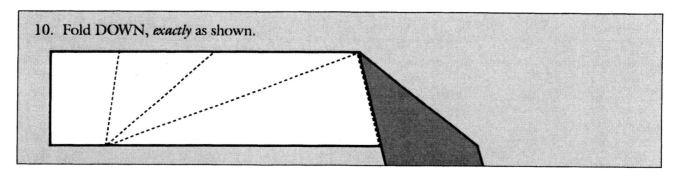

11. Unfold.

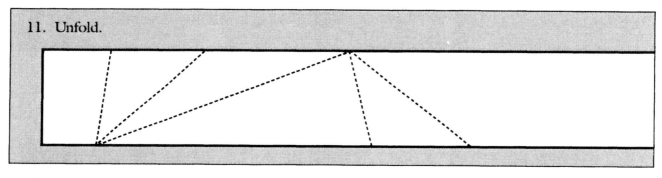

12. Fold DOWN, *exactly* as shown.

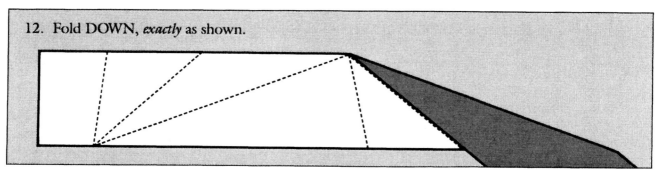

13. Unfold.

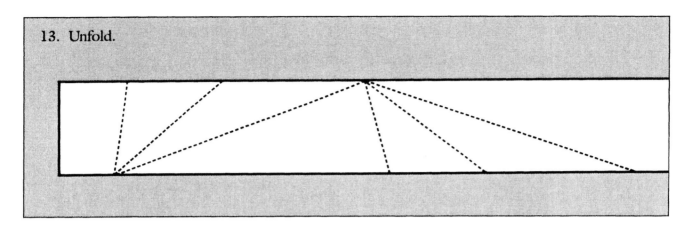

14. Fold UP, *exactly* as shown.

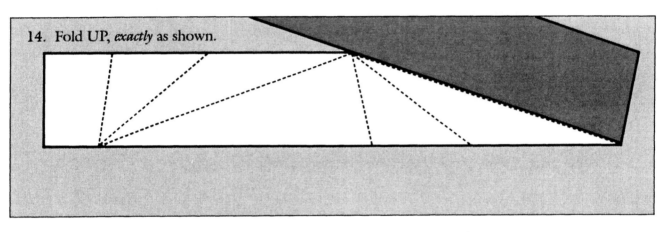

15. Now repeat steps 3 through 14. That is, continue folding UP, UP, UP, DOWN, DOWN, DOWN, UP, UP, UP, DOWN, DOWN, DOWN, ... (or U^3D^3). You will notice that, as you fold, the pattern of lines on the tape becomes more and more regular. First throw away the beginning part of the tape (say, the first 9 triangles). Now observe that the tape has *three* different kinds of lines. For simplicity let us call them *short, medium,* and *long* lines. Now experiment with your strip of tape and try creasing it on just one kind of line while leaving the other fold lines flat. You should crease *all* lines of the given kind. What do you think will happen? Make a guess and test it. After you have experimented and made the three different sizes of regular polygons that result, look at Figures 1.8–1.11 and see if the polygons you discovered are the same as ours.

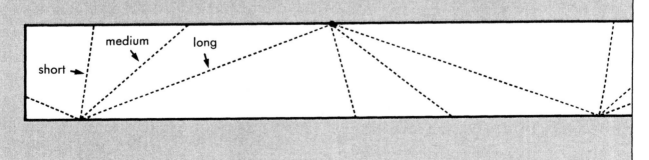

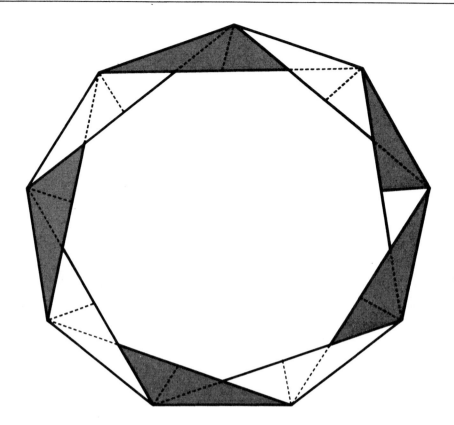

Figure **1.8**   A *long*-line 9-gon.

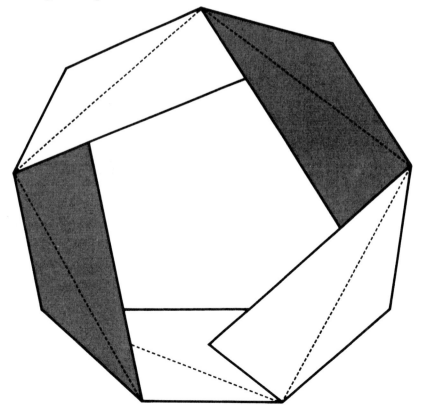

Figure **1.9**   A *medium*-line 9-gon.

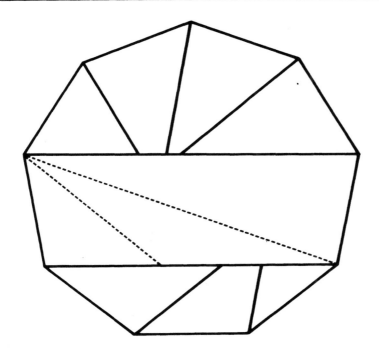

Figure **1.10**   A *short*-line 9-gon.

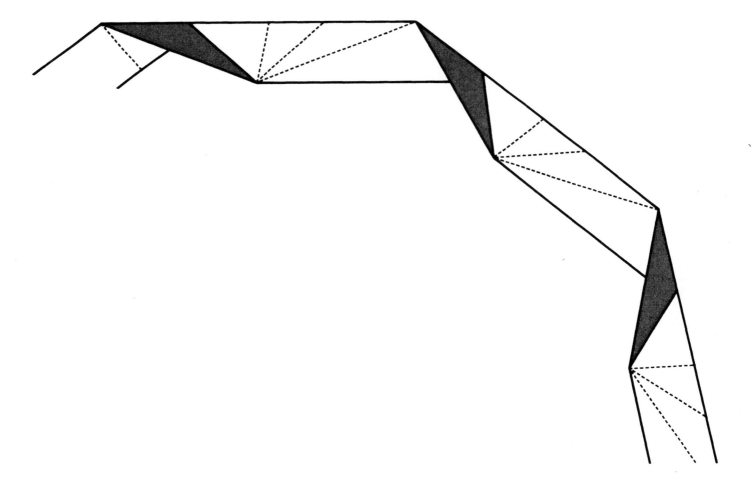

Figure **1.11**   Part of a 9-gon constructed by performing the F-A-T algorithm on *long* lines.

Now we know that the answer to our first question at the end of Section 1.4. is 9. So we need to find a reasonable general rule that will give 3, 5, and 9 for the first three entries. Let us now set out the evidence.

We return to the folding procedure in these three cases and write what we know in tabular form.

Case Number (and folding procedure)	Number of times angle (at top or bottom) was bisected	Angle made by longest fold line, as a fraction of original angle	Number of sides of resulting polygon
1 (U^1D^1)	1	$\frac{1}{2}$ (or $\frac{1}{2^1}$)	3
2 (U^2D^2)	2	$\frac{1}{4}$ (or $\frac{1}{2^2}$)	5
3 (U^3D^3)	3	$\frac{1}{8}$ (or $\frac{1}{2^3}$)	9

Ah hah! The numbers 2, 4, and 8 are very closely related to the sequence 3, 5, and 9. In fact, it is very easy to see that

$$3 = 2 + 1 = 2^1 + 1$$
$$5 = 4 + 1 = 2^2 + 1$$
$$9 = 8 + 1 = 2^3 + 1$$

So now we should suspect (and it turns out to be true) that if we fold U^nD^n, we can use that tape to construct regular convex $(2^n + 1)$-gons.

The next case in our list, namely $n = 4$, would yield $2^4 + 1 = 17$. Thus we now know how we may construct Gauss' beloved 17-gon. Simply fold a strip of tape using the U^4D^4 procedure and then apply the F-A-T algorithm (on the longest line) at 17 equally spaced vertices along the top of the tape. Alternatively, crease consistently along any one of the four different kinds of fold lines, while leaving the other fold lines flat.

In theory, we could use this method to construct the 33-gon (since $33 = 32 + 1 = 2^5 + 1$), the 65-gon (since $65 = 64 + 1 = 2^6 + 1$), the 129-gon, the 257-gon, and so on. In principle, we could even construct the 65537-gon (by folding $U^{16}D^{16}$). However, in practice, it is very difficult to fold either up or down in the prescribed manner more than four times.

Actually, we can do more with the U^3D^3 tape. You may have already noticed in your own investigations that it is possible to perform the F-A-T algorithm on the U^3D^3 tape along the medium-length lines. Figure 1.12 shows the *star* polygon that is obtained by this procedure. The notation we use for this polygon, that is, the $\{\frac{9}{2}\}$-gon, is the clever invention of the mathematician H.S.M. Coxeter. In this case, the denominator 2 indicates that the top edge of the tape visits successively every *second* vertex of some regular convex polygon, and the numerator 9 indicates that the regular convex polygon we are talking about has *nine* vertices (and hence nine sides). All our previous polygons could be denoted, if we wished, by this fractional notation, with 1 in the denominator, but as usual we follow the custom of just writing the numerator in this case. We may always regard a *convex* polygon as the special case of a star polygon in which the vertices are visited in their natural sequence.*

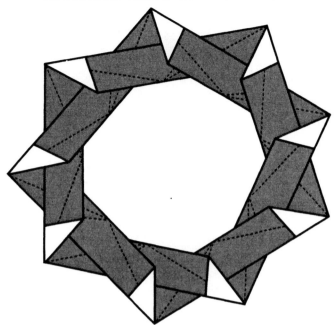

Figure **1.12**   A regular star $\{\frac{9}{2}\}$-gon, formed by performing the F-A-T algorithm on *medium* lines.

1.6   *Folding 7-gons*

So far we have given folding rules in which we treat the top and bottom of the tape in the same way. We have seen that this restriction will only allow us to construct $(2^n + 1)$-gons. Let us see whether we can construct a wider class of regular polygons if we use a different number of fold lines at the top and bottom of the tape. If so, it will surely be worth considering this more general folding procedure.

Let us start then with the simplest case, the D^2U^1 (or, equivalently, U^1D^2) procedure. We will show you in the last section of Chapter 2 *why* this produces the regular 7-gon.

Folding 7-gons

1. Begin with a straight strip of gummed tape.

---

*The strict definition of a polygon (or polygonal path) does not allow the sides of the polygon to cross each other. Thus a star polygon is not truly a polygon—but that doesn't mean it is not a beautiful geometrical figure of considerable mathematical interest.

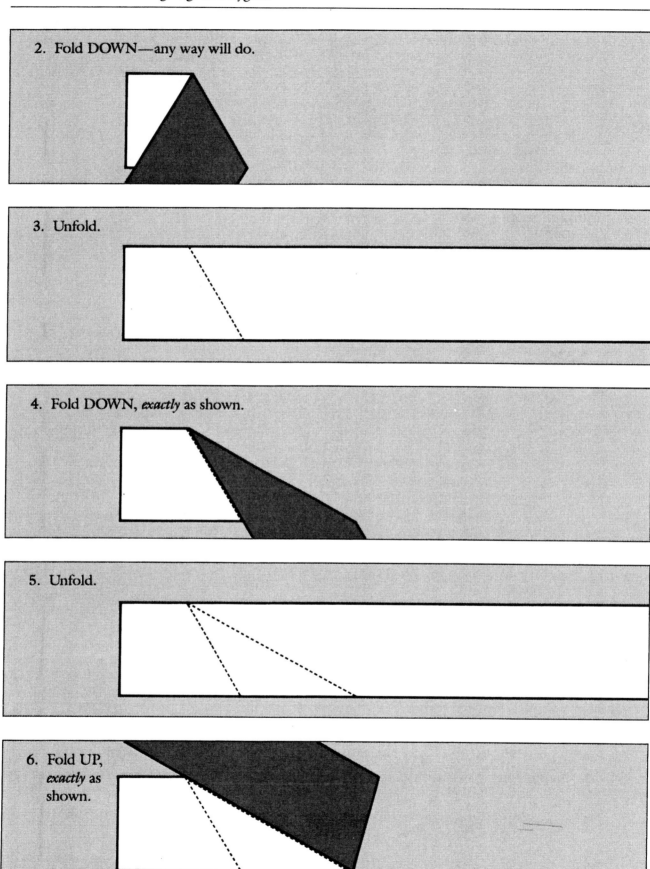

2. Fold DOWN—any way will do.

3. Unfold.

4. Fold DOWN, *exactly* as shown.

5. Unfold.

6. Fold UP, *exactly* as shown.

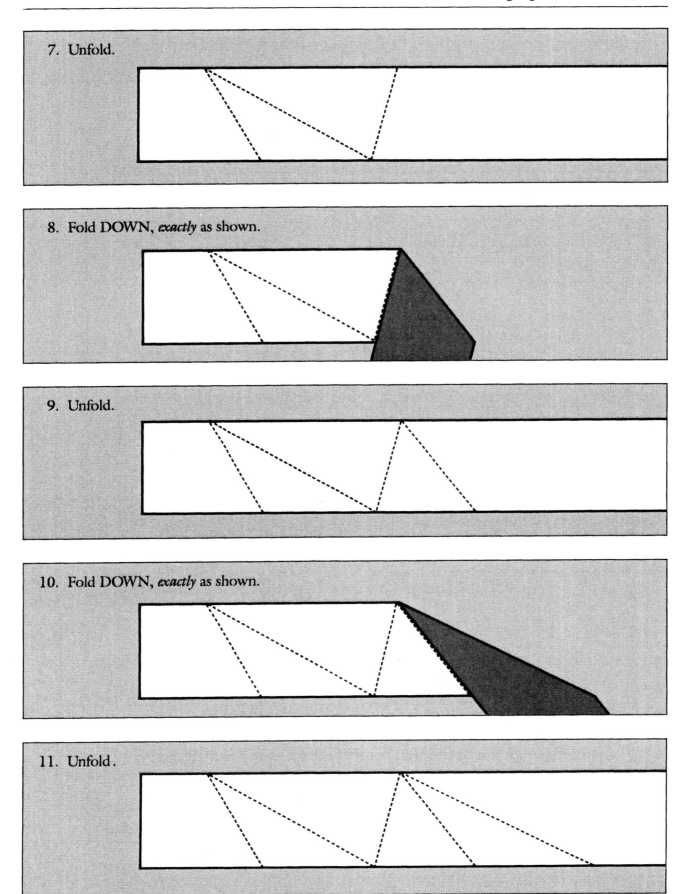

7. Unfold.

8. Fold DOWN, *exactly* as shown.

9. Unfold.

10. Fold DOWN, *exactly* as shown.

11. Unfold.

12. Fold UP, *exactly* as shown.

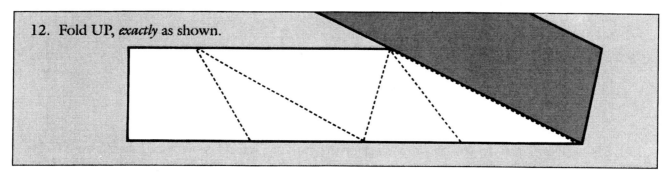

13. Unfold.

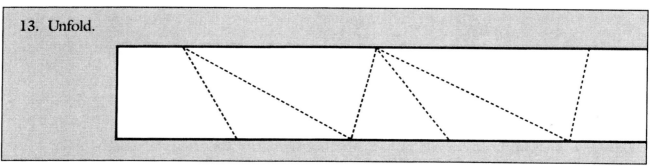

14. Now repeat steps 8 through 13. That is, continue folding DOWN, DOWN, UP, DOWN, DOWN, UP, . . . (or D^2U^1). As with our other folding procedures, the pattern of lines on the tape will become more and more regular as you continue to fold. Throw away the first part of the tape (say, the first 8 triangles). The remaining tape can now be used to construct star 7-gons, including the convex 7-gon, as indicated in Figure 1.13.

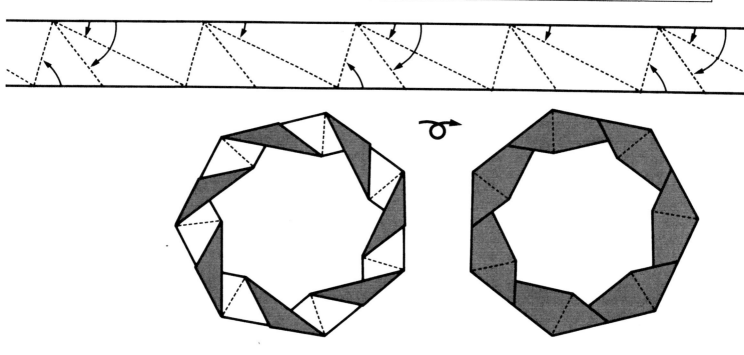

Figure **1.13**   (a) A convex 7-gon constructed by executing the F-A-T algorithm on successive *long* lines.

Figure **1.13** cont.

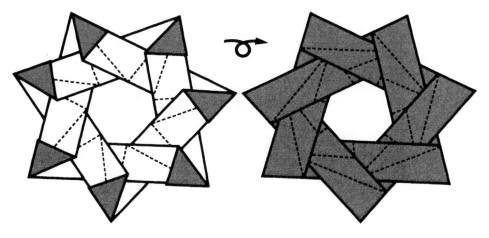

(b) A star $\{\frac{7}{2}\}$-gon constructed by executing the F-A-T algorithm on every other *medium* line starting from the *top* of the tape.

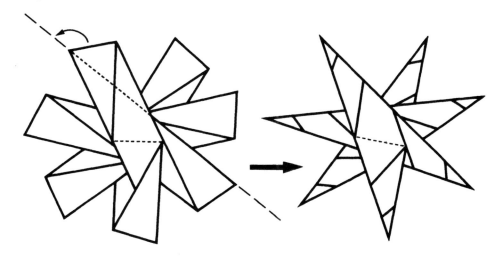

(c) A star $\{\frac{7}{3}\}$-gon constructed by executing the F-A-T algorithm on every other *short* line starting on the *bottom* of the tape.

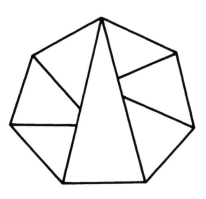

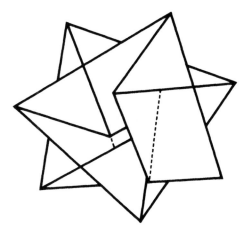

(d) A 7-gon formed by creasing along all *long* and *short* lines, leaving the medium lines flat.

(e) A star $\{\frac{7}{2}\}$-gon formed by creasing along all *short* and *medium* lines, leaving the long lines flat.

1.7 Folding Squares and 8-gons

The constructions in this section are *exact* (but only in theory!). Consequently, when you follow the folding instructions you will be able to use all the tape without throwing away any portion at the beginning. There are many methods for folding squares and octagons. You probably already know some of them.

Since it is more exciting to try to discover things for yourself than to follow what someone else has done, we suggest that you take a strip of gummed tape and try to fold some squares or octagons before looking at our instructions. You may discover some pretty ones we have not thought of. After you have experimented with your own ideas, then look at our instructions.

In the constructions for squares and 8-gons (octagons) that we describe, we have selected—from the many possibilities—procedures that enable us to use the F-A-T algorithm and also to build some very interesting related models. Our construction for a square may surprise you, since you may have been expecting to fold a strip to look like this:

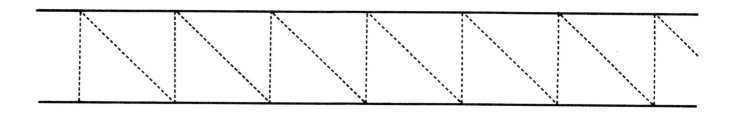

You may, in fact, wish to fold a strip like this and execute the F-A-T algorithm on each of the long lines. The reason we chose the following instructions for folding a square is that the strip can then be used to construct a very interesting mathematical toy, the 8-flexagon of Section 3.3, and a very attractive polyhedron, the diagonal cube of Section 9.3.

As we have said, you may try any construction that appeals to you. We certainly want you to try out your own ideas and to experiment with your own constructions. Then, after you've tried to construct your own squares and octagons, you may want to look at the following pages.

Folding Squares

1. Start with a straight strip of gummed tape.

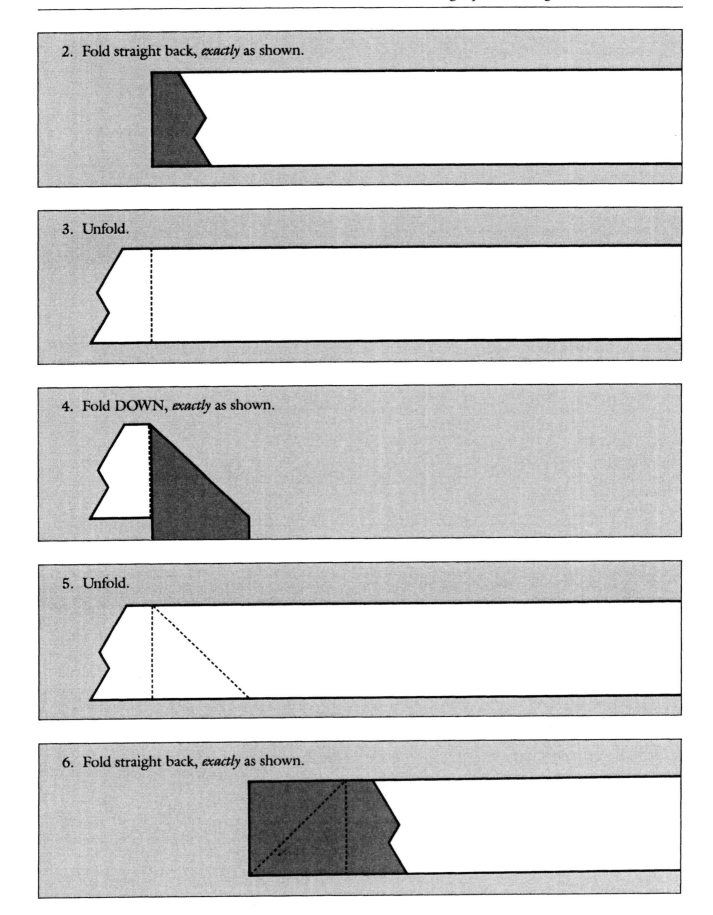

2. Fold straight back, *exactly* as shown.

3. Unfold.

4. Fold DOWN, *exactly* as shown.

5. Unfold.

6. Fold straight back, *exactly* as shown.

7. Unfold.

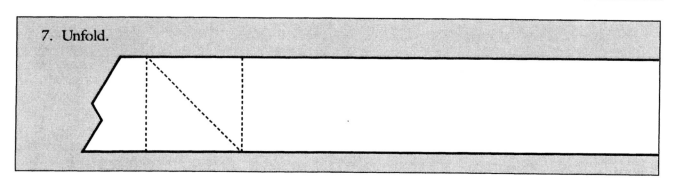

8. Fold UP, *exactly* as shown.

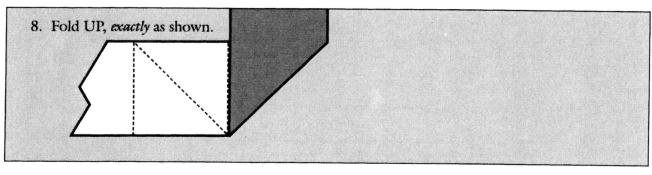

9. Unfold.

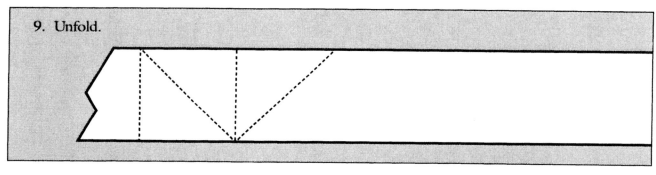

10. Fold straight back, *exactly* as shown.

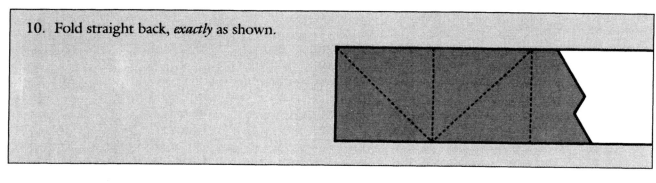

11. Unfold.

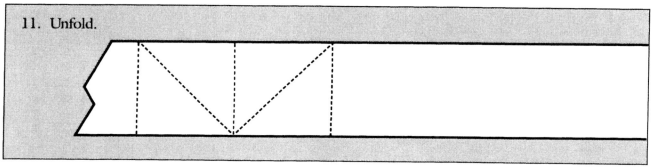

12. Continue folding in this way until you get a long strip of triangles that look just like this. These triangles will be regular from the beginning. You won't need to throw any of them away!

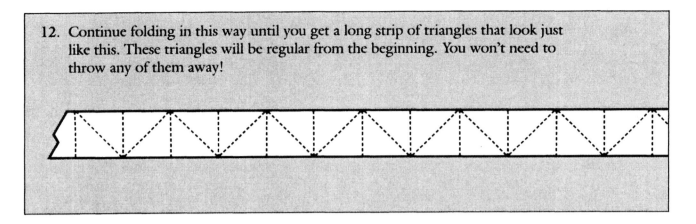

Figures 1.14 and 1.15 show two different ways of folding a square.

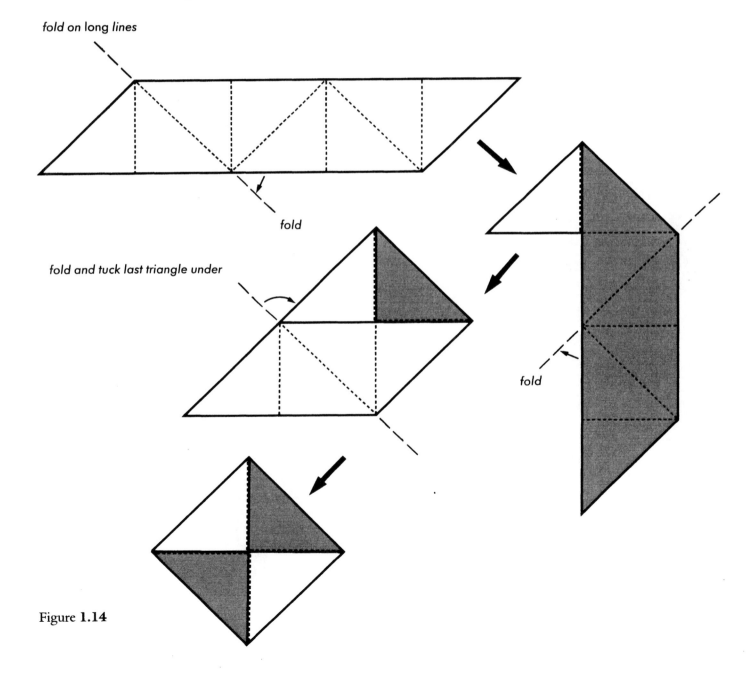

fold on long lines

fold

fold and tuck last triangle under

fold

Figure **1.14**

Begin with a strip of 17 triangles.

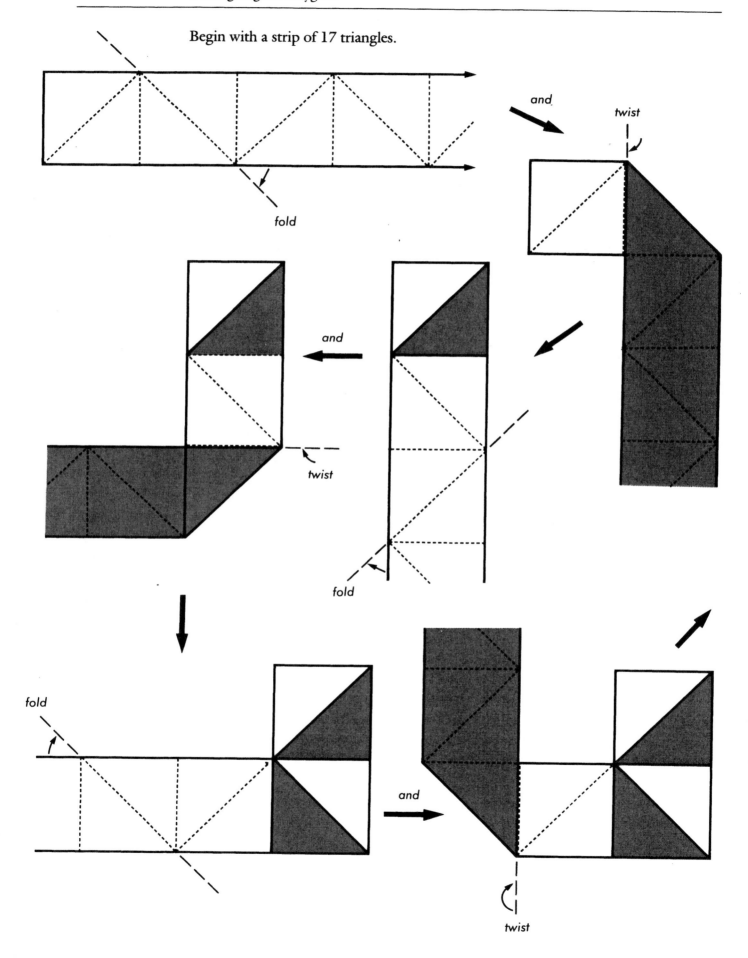

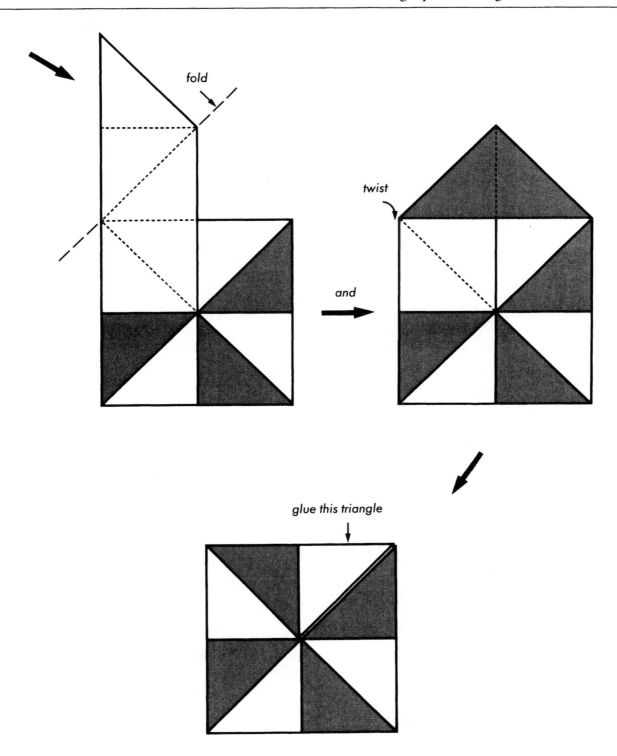

fold

twist

and

glue this triangle

Figure **1.15**

Actually, we have described something far more interesting than a mere square in Figure 1.15. What we have here is another example of a flexagon. Since it is a square, it is sometimes called a *tetraflexagon*. We will call it an *8-flexagon* because its surface consists of 8 triangles. See Section 3.3 for details about flexing your new toy.

Folding 8-gons

1. Begin with a straight strip of gummed tape.

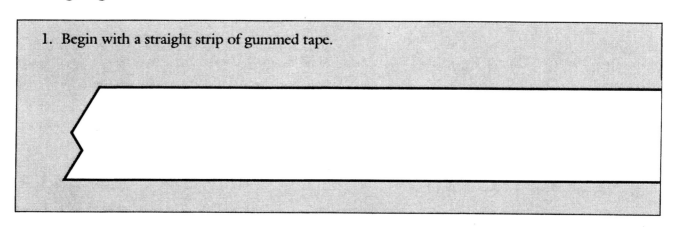

2. Fold straight back, *exactly* as shown.

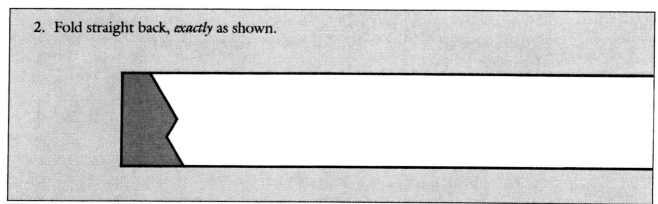

3. Unfold.

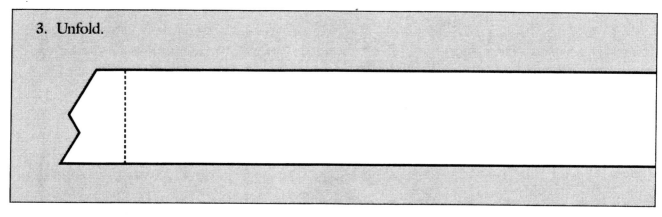

4. Fold DOWN, *exactly* as shown.

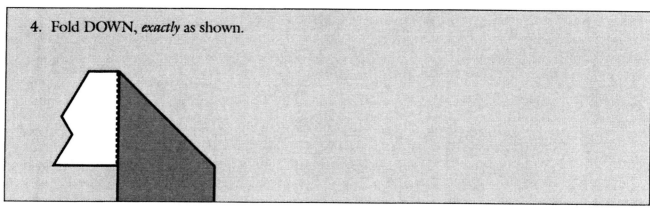

5. Unfold.

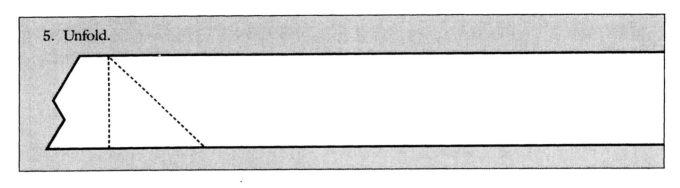

6. Fold DOWN, *exactly* as shown.

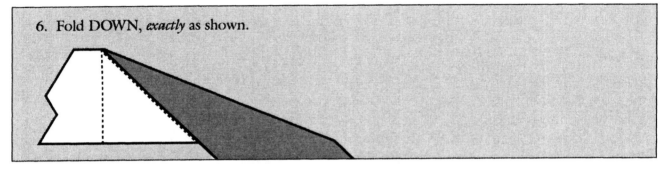

7. Unfold.

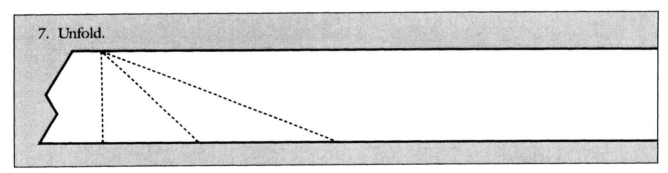

8. Fold straight back, *exactly* as shown.

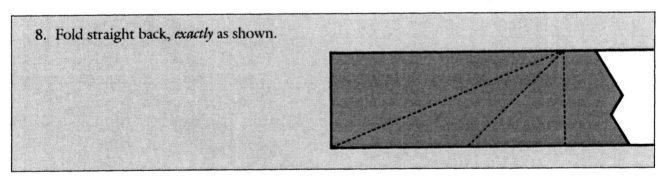

9. Unfold.

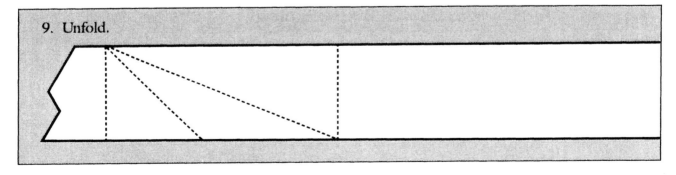

10. Continue folding, repeating steps 4 through 9. You will get a strip of tape that looks like the one in Figure 1.16. You can use this strip to fold a regular 8-gon by executing the F-A-T algorithm along each long line, or by creasing on each long and short line and leaving the medium-length lines flat. You may discover other ways to fold 8-gons on your own. See also Figure 1.17.

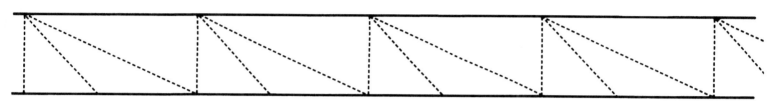

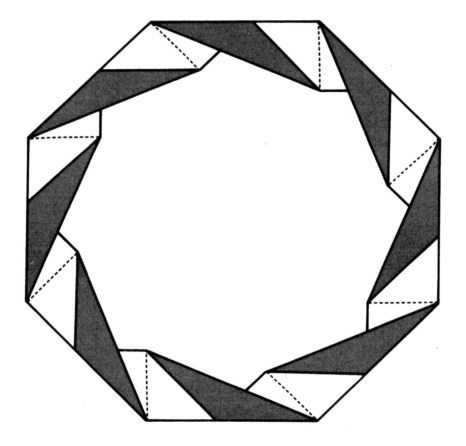

Figure 1.16   An 8-gon constructed by executing the F-A-T algorithm on *long* lines.

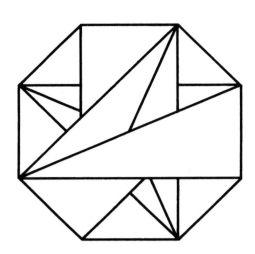

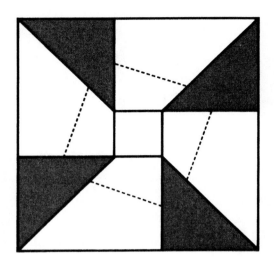

Figure 1.17

(a) An 8-gon formed by creasing on *long* and *short* lines, leaving the medium lines flat.

(b) A 4-gon formed by executing the F-A-T algorithm on *medium* lines.

Just as we introduced secondary fold lines into the D^1U^1-folded tape to produce 6-gons instead of 3-gons, and into the D^2U^2-folded tape to produce 10-gons instead of 5-gons, so we could introduce secondary fold lines into any tape from which we constructed squares in order to produce 8-gons. However, in our opinion the resulting 8-gon is not quite as attractive as the one constructed by the method we have just described.

2 The Mathematics of Paper Folding

This chapter is included for those of you who would like to know just what can be done by paper folding, using the F-A-T algorithm, and why it works. Many of you may prefer to go straight on to Chapter 3 and subsequent chapters, rather than take the time now to master this theoretical material. That would be a perfectly reasonable and understandable thing to do. However, we do not recommend that you completely ignore the mathematical basis of the practical work you will be doing with polygons and polyhedra. We do want you to understand why the constructions we describe in this book work. So this chapter and Chapter 12 are, in our view, very important chapters.

For those still with us, we're glad you're here! We now begin.

2.1. Angles

Let us look again at our construction of a regular 9-gon by the D^3U^3 folding procedure. We are going to explain to you now why it works, and we start with some important facts about angles. Suppose that we have any 9-gon:

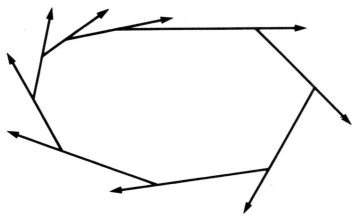

We have extended the sides of the polygon so that you may see the *exterior angles*. We make the following claim:

For any polygon, the sum of the exterior angles is 360°.

To see this, we use our "argument by myopia"! A shortsighted person, looking at our polygon from a considerable distance, would see this:

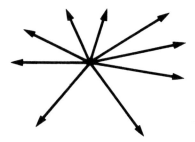

This person would immediately conclude the truth of our claim; one moves through one complete revolution in going around the polygon. Of course, this has nothing to do with the polygon having 9 sides and would work for any polygon.

Now suppose that the polygon is regular. Then all the exterior angles are equal, so each is $\frac{360°}{p}$, where p is the number of vertices (which is, of course, the same as the number of sides). Now let's look at the *interior* angles. Each is the supplement of the corresponding exterior angle, so we get the following result:

Each exterior angle of a regular p-gon is $\frac{360°}{p}$.

Each interior angle of a regular p-gon is $(180 - \frac{360}{p})°$.

So, for our regular 9-gon, the exterior angles are 40° and the interior angles are 140°. It thus follows that if we have marked 9 points at regularly spaced intervals along the top of the tape, and if we have a procedure for turning the tape through 40° at each of these points, then the top of the tape will form itself into a regular 9-gon. It remains, of course, to convince you that our folding procedure U^3D^3 (which is equivalent to D^3U^3), combined with the F-A-T algorithm, does turn the top edge of the tape through 40° at regularly spaced points, but we will do this a little later.

Before moving on, however, let's look at a star polygon, say, the $\{\frac{7}{3}\}$-polygon, or $\{\frac{7}{3}\}$-gon:

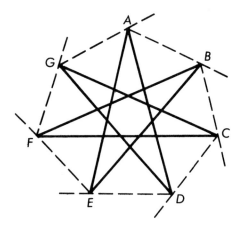

We have agreed that to construct a regular convex 7-gon, the tape must turn, at regularly spaced intervals, through an angle of $\frac{360°}{7}$. If the vertices of our star polygon are indeed vertices of a regular 7-gon, then from A to D we have turned 3 times through an angle of $\frac{360°}{7}$, that is, we have turned through an angle of $(\frac{3}{7} \times 360)°$. Thus we see that to fold a $\{\frac{7}{3}\}$-gon, we need to be able to turn the top of the tape, at regularly spaced points, through an angle of $(\frac{3}{7} \times 360)°$.

We now remark that if—somehow—we have put fold lines into the tape making angles of d degrees with the top of the tape, at regular intervals along the top of the tape, then the F-A-T algorithm allows us to turn the top of the tape, at these points, through an angle of $2d$ degrees:

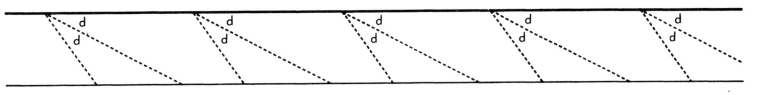

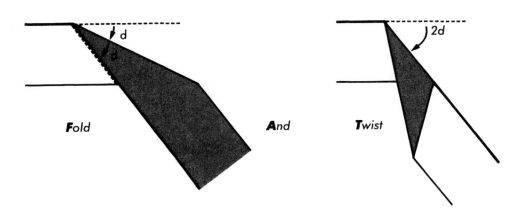

Fold   **A**nd   **T**wist

So we can obtain a $\{\frac{7}{3}\}$-gon if we can put fold lines into the tape, at regular intervals, making angles of $(\frac{3}{7} \times 180)°$. Quite generally, we have the following result:

> If we can put fold lines into the tape, at regular intervals, making an angle of $(\frac{a}{b} \times 180)°$ with the top of the tape, then the F-A-T algorithm enables us to construct a regular $\{\frac{b}{a}\}$-gon.

In fact, the main problem is to devise ways of putting in fold lines making angles of $(\frac{a}{b} \times 180)°$ *where a and b are both odd*. For there are fairly obvious ways, given such fold lines, of introducing secondary fold lines to make angles of

$$\frac{2a}{b} \times 180°, \frac{4a}{b} \times 180°, ..., \frac{a}{2b} \times 180°, \frac{a}{4b} \times 180°, ...$$

We already described a few such secondary fold lines in Chapter 1 when we discussed the construction of 6-gons, 12-gons, and 10-gons. So the real problem is concerned with the case of constructing $\{\frac{b}{a}\}$-gons with b and a both odd.

There is a further small point to be made here. In discussing $\{\frac{b}{a}\}$-gons, we obviously want a and b to have no common factor (why?); but we also want to restrict ourselves to the case where $a < \frac{b}{2}$. This is because, for example, a $\{\frac{7}{4}\}$-gon is really the same thing as a $\{\frac{7}{3}\}$-gon (why?).

We can sum up our discussion so far as follows:

We know how to construct all regular star polygons, once we know how to construct regular $\{\frac{b}{a}\}$-gons with b and a both odd. Thus the problem is to put fold lines in the tape making angles of $(\frac{a}{b} \times 180)°$ with the top of the tape, where b and a are odd with $a < \frac{b}{2}$.

So let us now discuss this problem. In this section we are content to make a few crucial preliminary remarks.

We first consider a special case. We have seen that to construct a regular 7-gon, we must succeed in putting in fold lines at equally spaced intervals, making angles of $\frac{180°}{7}$ with the top of the tape. Now suppose that, somehow, we have put in *one* such fold line, as shown:

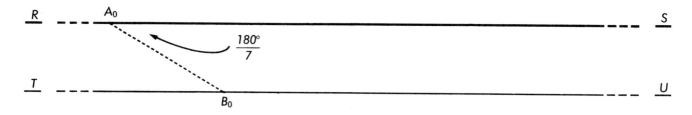

Because RS and TU are parallel lines, the angles SA_0B_0 and A_0B_0U are *supplementary* (that is, their sum is 180°). Thus the angle A_0B_0U is $(\frac{6}{7} \times 180)°$. Let us put in an UP fold line (exactly as we did in Chapter 1) to bisect this angle:

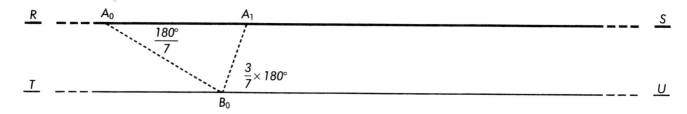

Again, by the rule of supplementary angles, the angle SA_1B_0 is $(\frac{4}{7} \times 180)°$. If we fold DOWN once (exactly as we did in Chapter 1), we bisect this angle; if we fold DOWN again, this angle is divided by 4. The resulting angle is $(\frac{1}{7} \times 180)°$—exactly what we want!

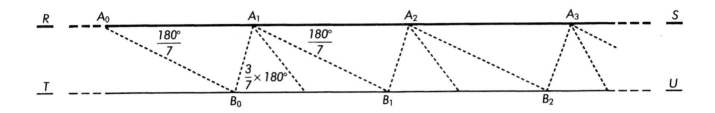

Thus we may repeat the procedure. It is obvious that the points $A_0, A_1, A_2, \ldots$ are appearing at equally spaced intervals along the top of this tape. Likewise, the points $B_0, B_1, B_2, \ldots$ are appearing at equally spaced intervals along the bottom of the tape. Moreover, the folding procedure we have described is exactly the D^2U^1 procedure we introduced earlier to construct a regular 7-gon. Notice, too, that if we use the bottom of the tape instead of the top, we construct the regular $\{\frac{7}{3}\}$-gon, and that if we use the medium fold lines at the top, we construct the regular $\{\frac{7}{2}\}$-gon.

Everything seems then to hinge on whether we can put in that first fold line A_0B_0. However, we show you in the next section that this is no barrier to achieving our construction provided we are content with an arbitrarily good approximation to our desired regular polygon. Here we consider the question of just how general our method is. **Would it work for any $\{\frac{b}{a}\}$-gon, with b and a odd?**

Before we attempt to answer this question, we'd better be more precise as to what we mean by "our method." This consists of supposing a correct initial fold line put into the tape making an angle of $(\frac{a}{b} \times 180)°$ with the top of the tape, passing to the supplementary angle at the bottom of the tape, which will have the form $(\frac{b-a}{b} \times 180)°$, and then halving this angle by bisection until it takes the form $(\frac{a'}{b} \times 180)°$, *with a' again odd*. Then we go to the supplementary angle $(\frac{b-a'}{b} \times 180)°$ at the top of the tape and again halve this angle by bisection until it takes the form $(\frac{a''}{b} \times 180)°$ with a'' odd; and so on. So now we repeat the question: Does the method always work?

The answer is, yes and no! It is "no" in the sense that we certainly cannot construct *any* such $\{\frac{b}{a}\}$-gon if we insist that the folding instructions be of the simple kind, "fold down m times, fold up n times (D^mU^n)." In Section 2.3 we discuss which ones we can get that way. However, the answer is "yes" in the sense that we can construct any such $\{\frac{b}{a}\}$-gon if we allow a rather more complicated—but equally systematic—folding procedure.

Let us illustrate this last remark by considering the regular 11-gon. As we know, making a regular 11-gon necessitates our putting in fold lines making angles of $\frac{180°}{11}$ with the top of the tape at regular intervals. Imitating our procedure with the 7-gon, we suppose one such fold line A_0B_0 put in. We may now tabulate our subsequent moves, as follows:

Angle A_0B_0U is $(\frac{10}{11} \times 180)°$.

Bisect it to get angle A_1B_0U of $(\frac{5}{11} \times 180)°$.

Angle SA_1B_0 is $(\frac{6}{11} \times 180)°$.

Bisect it to get angle SA_1B_1 of $(\frac{3}{11} \times 180)°$.

Angle A_1B_1U is $(\frac{8}{11} \times 180)°$.

Bisect it 3 times to get angle A_2B_1U of $(\frac{1}{11} \times 180)°$.

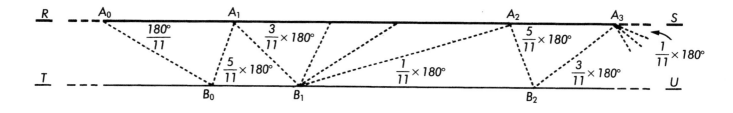

Notice that we have now achieved the angle we want, but it has appeared at the bottom of the tape. But that's no problem—repeat the sequence! Thus we get angle SA_2B_2 of $(\frac{5}{11} \times 180)°$, angle A_3B_2U of $(\frac{3}{11} \times 180)°$, and angle SA_3B_3 of $(\frac{1}{11} \times 180)°$. Continuing this way, fold lines appear at the top of the tape at equally spaced points $A_0, A_3, A_6, A_9, ...$, all making angles of $\frac{180°}{11}$ with the top of the tape. The folding instructions may be written $D^3U^1D^1U^3D^1U^1$. Thus the folding instructions are, as we claimed, more complicated but just as systematic; and they can be deduced *by simple arithmetic* from the odd numbers b and a involved—in this case, $b = 11$ and $a = 1$. It may be shown that *this arithmetic always works*, that is, for *any* odd numbers b and a for which we wish to construct a regular $\{\frac{b}{a}\}$-gon, the arithmetic will give us a folding procedure for doing so, *once the initial fold line has been put into the tape*. The original angle, $(\frac{a}{b} \times 180)°$, may reappear first at the top or the bottom of the tape. If (as in our example) it reappears first at the bottom, just repeat the folding sequence, so that its next appearance is at the top.

2.2 Approximation and Convergence

We seem to have reduced our problem to that of making the initial fold in the tape—but is that not an appallingly difficult thing to do? We now show that it is, in fact, ridiculously easy provided we are content to have our resulting polygons be arbitrarily good approximations. (In a deep philosophical sense it would be an impossible thing to do if we insisted on absolute accuracy!)

In fact, we show that *our initial fold can be as inaccurate as we please!* Let us explain this apparently strange remark. We revert to our discussion of the regular 7-gon, and we now suppose that our initial fold line $A_0 B_0$ actually made an angle of $(\frac{180}{7} + E)°$ with the top of the tape. Here we think of E as an *error*, which may, of course, be positive or negative. Following through the arithmetic, we see that $B_0 A_1$ makes an angle of $(\frac{3}{7} \times 180 - \frac{E}{2})°$ with the bottom of the tape; and $A_1 B_1$ makes an angle of $(\frac{180}{7} + \frac{E}{8})°$ with the top of the tape. Thus the error has been reduced to $\frac{E}{8}$. If we look at $A_2 B_2$ the angle will be, by the same argument, $(\frac{180°}{7} + \frac{E}{64})°$, and so on. Now our initial error E would certainly be no bigger than $\pm 20°$ if we just guessed, and it is very unlikely to be bigger than $\pm 10°$. Thus, in the worst case, the error in the direction of $A_2 B_2$ is less than $20'$, and the error in the direction of $A_4 B_4$ is less than $20''$. So we may regard $A_4 B_4$ as our initial fold line! This is the reason for the instruction given to start folding and throw away the first (that is, the left-hand) part of the folded tape. It is, of course, plain that as the direction of $A_n B_n$ gets closer and closer to $\frac{180°}{7}$, the distance between successive points A_n and A_{n+1} on the top of the tape undergoes less and less change as we move to the right along the tape.

The justification for our procedure works in general. Suppose we had to introduce k fold lines in passing from A_0 to the next point A_r on the top of the tape at which, ideally, the fold line is parallel to the one at A_0 (thus, for the 7-gon, $k = 3$; for the 11-gon, $k = 10$). (Let us call this point A_r.) Then it is not difficult to see that if the initial error was E, the error in the direction of the fold line at A_r is $\frac{E}{2^k}$. So, for our 11-gon, an initial error of $10°$ is immediately reduced to an error of less than $0.01°$. Here we have a beautiful example of a (rapidly) *converging process*. The sequence of angles at these successive points on the top of the tape converges very quickly to the angle we want; and, as the angles converge, so does the distance between successive points. Of course, the *actual* distance to which this sequence converges, which will be the distance between successive vertices on the polygonal path we construct, depends on the width of our tape.

2.3 The Simple Folding Procedure $D^m U^n$

In this section we answer the question: **What star polygons can we fold if we confine attention to the folding procedure $D^m U^n$, where we fold DOWN m times at every vertex at the top of the tape and UP n times at every vertex at the bottom of the tape?***

---

*Of course, $D^m U^n$ is the same as $U^n D^m$. Henceforth in this section we will write the "D" before the "U."

Suppose this folding procedure yields an angle of $x°$ at the top and an angle of $y°$ at the bottom:

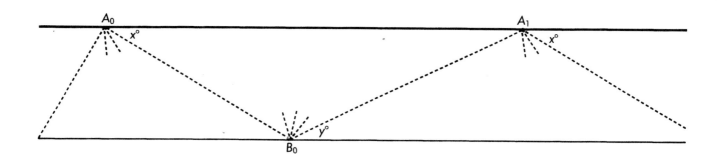

Then, if we look at B_0, we see that the supplement of x was halved (bisected) n times to produce y, so that

$$x + 2^n y = 180$$

If, likewise, we look at A_1, we see that the supplement of y was halved (bisected) m times to produce x, so that

$$y + 2^m x = 180$$

We have a pair of simultaneous equations in x and y to solve. Let us solve by eliminating y; we find

$$x + 2^n (180 - 2^m x) = 180$$
$$(2^{m+n} - 1)x = (2^n - 1)180$$
$$x = \frac{2^n - 1}{2^{m+n} - 1}180$$

Similarly,

$$y = \frac{2^m - 1}{2^{m+n} - 1}180$$

We have proved the following:

> The $D^m U^n$ procedure allows us to construct a $\{\frac{2^{m+n} - 1}{2^n - 1}\}$-gon using the top of the tape and a $\{\frac{2^{m+n} - 1}{2^m - 1}\}$-gon using the bottom of the tape.

To see that this fits with what we know, let us first consider the procedures $D^1 U^1$, $D^2 U^2$, and $D^3 U^3$. Of course, with these procedures, it does not matter whether we use the top or bottom of the tape. (Why?) With $D^1 U^1$ we make a $\{\frac{2^2 - 1}{2 - 1}\}$-gon, that is, a 3-gon; with $D^2 U^2$ we make a $\{\frac{2^4 - 1}{2^2 - 1}\}$-gon, that is, a 5-gon; with $D^3 U^3$ we make a $\{\frac{2^6 - 1}{2^3 - 1}\}$-gon, that is, a 9-gon; with $D^n U^n$ we make a $\{\frac{2^{2n} - 1}{2^n - 1}\}$-gon, that is, a $(2^n + 1)$-gon. All this you saw, or guessed, much earlier.

Now let us look at the $D^2 U^1$ procedure. Using the top of the tape we construct a $\{\frac{2^3 - 1}{2 - 1}\}$-gon, that is, a 7-gon; and, at the bottom of the tape, we have fold lines enabling us to construct a $\{\frac{2^3 - 1}{2^2 - 1}\}$-gon, that is, a $\{\frac{7}{3}\}$-gon. This, too, confirms earlier results.

What *convex* polygons can we construct by our simple procedure? The answer is plain—we can construct a convex N-gon if N has the form $\frac{2^{m+n} - 1}{2^n - 1}$. But is this answer helpful? How can we recognize if a given number N has this form? It turns out that this question admits a very striking answer, which we give in two parts.

The fraction $\frac{2^{m+n} - 1}{2^n - 1}$ is an integer if and only if n is a factor of m.

Let us write N in base 2. Then N has the form $\frac{2^{m+n} - 1}{2^n - 1}$ if and only if, in base 2,

$$N = \underbrace{10 \dots 0}\ \underbrace{10 \dots 0}\ \dots\ \underbrace{10 \dots 0}\ 1$$

where $10 \dots 0$ repeats $\frac{m}{n}$ times and consists of 1 followed by $(n - 1)$ zeros.
We call $10 \dots 0$ the repeating pattern of N.

Notice that the second part not only enables us to recognize the numbers N that have the form $\frac{2^{m+n} - 1}{2^n - 1}$, but also tells us what m and n are in terms of N, thus enabling us to determine just which $D^m U^n$ folding procedure produces a specific convex N-gon. Let us look at some examples.

Example 1 Let $N = 7$. In base 2, $7 = 111$. Thus a 7-gon may be constructed by folding $D^m U^n$ where $n = 1$ (since there are no zeros in the repeating pattern) and $\frac{m}{n} = 2$, so that $m = 2$ (remember that the final 1 is *not* part of the repeating pattern).

Example 2 Let $N = 21$. In base 2, $21 = 10101$. Thus a 21-gon may be constructed by folding $D^m U^n$ where $n = 2$ (since there is one zero in the repeating pattern) and $\frac{m}{n} = 2$, so that $m = 4$ (since 10 appears twice).

Example 3 Let $N = 11$. In base 2, $11 = 1101$. This does *not* have the right pattern, so we cannot construct an 11-gon by any $D^m U^n$ folding procedure. Indeed, we saw that we needed a more complicated procedure.

Finally: **What star polygons can we construct by some $D^m U^n$ procedure?** The question is, naturally, more difficult, but the answer is very rewarding. Let us call those numbers N such that we can construct convex N-gons by (some) $D^m U^n$ procedure the *folding numbers*. Thus a folding number is a number of the form

$$\frac{2^{m+n} - 1}{2^n - 1}$$

where n is a factor of m. Then we may prove the following:

> The star polygons that we can construct by our simple folding procedure are those $\{\frac{N}{a}\}$-gons where N is a folding number and a is a *prime section* of N.

It remains to explain what we mean by a *prime section* of N. To understand this, we revert to our representation of N in base 2,

$$N = \underbrace{10 \dots 0}\ \underbrace{10 \dots 0}\ \dots\ \underbrace{10 \dots 0}\ 1$$

Then a *section* of N is a number represented in base 2 by the right-hand portion of N, starting at some 1. Thus, for example,

$$\text{if } N = 1001001001 = 585$$
$$\text{then } a = \quad\ \ 1001001 = \quad 73 \text{ is a section.}$$

Of course, for it to be reasonable to talk of an $\{\frac{N}{a}\}$-gon, we need $a<\frac{N}{2}$ and, further, a and N should not have any common factor (except 1). The inequality $a<\frac{N}{2}$ is *always* true if a is a *proper section* of N (that is, if a is a section different from N itself). The following lovely property of folding numbers tells us whether a and N have a common factor.

> If N is a folding number with p 1's when written in base 2, and if a is a section of N with q 1's when written in base 2, then N and a have no common factor precisely when p and q have no common factor. In that case, we say that a is a *prime section* of N.

Example 4 Let $N = 101010101010101 = 21845$ (here $p = 8$).
  Then the sections of N are as follows:

$$
\begin{array}{rrr}
q=1 & 1 = & 1 \\
2 & 101 = & 5 \\
3 & 10101 = & 21 \\
4 & 1010101 = & 85 \\
5 & 101010101 = & 341 \\
6 & 10101010101 = & 1365 \\
7 & 1010101010101 = & 5461 \\
\end{array}
$$

As predicted, if $q = 2$, 4, or 6, a has a factor in common with 21845 (obviously, 5 is such a factor); you may verify that if $q = 1$, 3, 5, or 7, then a has no factor in common with 21845, that is, a is a prime section of 21845.

You now know (though we don't advise you to verify this by actually folding a strip of paper!) that you could, using some simple $D^m U^n$ procedure, construct the convex 21845-gon and the star

$$\{\tfrac{21845}{21}\}\text{-},\ \{\tfrac{21845}{341}\}\text{-},\ \{\tfrac{21845}{5461}\}\text{-gons!!}$$

You should now have one final question to ask: **If N is a given folding number and a is a section of N that has no factors in common with N, how do you choose m and n so that the $D^m U^n$ procedure enables you to construct a star $\{\frac{N}{a}\}$-gon?** The answer is again quite precise. Note first that if a *convex* N-gon is constructed by using $D^{m_1} U^{n_1}$ and if a convex a-gon is constructed by using $D^{m_2} U^{n_2}$, then $n_1 = n_2$. (Why? Remember that a is a *section* of N.) Then we may prove the following:

If a convex N-gon is constructed by using $D^{m_1} U^{n_1}$ and a convex a-gon is made by using $D^{m_2} U^{n_2}$, then $n_1 = n_2$ and a star $\{\frac{N}{a}\}$-gon is constructed by using $D^m U^n$, where $m = m_1 - m_2$, $n = n_1 + m_2$.

Example 5 We know that a convex 7-gon is constructed by using $D^2 U^1$. In base 2, $7 = 111$ and $3 = 11$, so 3 is a section of 7. Moreover, we know that a convex 3-gon is constructed by using $D^1 U^1$. We have already verified that, in this case, $n_1 = n_2 \ (= 1)$, $m_1 = 2$, and $m_2 = 1$. Thus a star $\{\frac{7}{3}\}$-gon is constructed by using $D^1 U^2$. We also knew this, since we had observed that we could construct a $\{\frac{7}{3}\}$-gon by using the $D^2 U^1$ procedure and then using the *bottom* of the tape.

For those keen to read further into the mathematical mysteries of folding paper, we suggest you consult our references.

Remark

Throughout this chapter we have measured angles in degrees. There is another angular measure called *radian* measure. If an angle subtends an arc of length l on the circle of unit radius, we say that l is the radian measure of that angle. Thus

$$180° = \pi \text{ radians}$$

You will see that our results are really better expressed using radians rather than degrees.

REFERENCES

Hilton, Peter, and Jean Pedersen. "Approximating Any Regular Polygon by Folding Paper: An Interplay of Geometry, Analysis and Number Theory." *Mathematics Magazine*, 56 (1983), pp. 141–155.

Hilton, Peter, and Jean Pedersen. "Folding Regular Star Polygons and Number Theory." *Mathematical Intelligencer*, 7, no. 1 (1985), pp. 15–26.

Hilton, Peter, and Jean Pedersen. "Geometry in Practice and Numbers in Theory," *Undergraduate Journal of Mathematics*, Monograph (1987).

3   Constructing Flexagons

Required Materials

☐ Strips (or a roll) of gummed mailing tape or adding machine tape about 1½ in. wide. The glue on the gummed tape should be the type that needs to be moistened to become sticky. Don't try to use tape that is sticky to the touch when dry.

☐ White glue or paper clips, but only if your folding tape is not gummed.

3.1   Basic Instructions

In this chapter we describe how to construct some special polygons that change their appearances when they are manipulated in certain ways.

In general, we refer to these configurations as *N-flexagons*, where *N* indicates the number of congruent triangles surrounding the center of the regular polygon formed by the constructed figure. We should point out that in the case of the 8-flexagons, the bounding polygon of the construction has 4 sides, not 8 as you might expect. For this reason the 8-flexagon is sometimes referred to as a *tetraflexagon* (see Section 1.7).

We describe in detail how to construct and flex two special cases for each of the values $N = 6$ and $N = 8$. We also refine the nomenclature by adding another number at the beginning of the name to indicate how many different complete "faces," of N triangles each, that the model may present when it is flexed. This nomenclature will be illustrated and described in detail at the appropriate time—and, for those of you who like to use big words, we will also give you the names that utilize Greek prefixes instead of the numbers.

We first discuss the 6-flexagons (already well known in the literature as *hexaflexagons*) in Section 3.2. We do this because they are the easiest flexagons to manipulate. Our idea is that you are likely to find it more pleasant to develop your flexing skills on the 6-flexagons, and that by doing this you will be better prepared to appreciate the considerably more complicated 8-flexagons described in Section 3.3. Of course, in keeping with the spirit of this entire book, we suggest variations or references along the way so that you can construct (or maybe even *invent*) other flexagons on your own.

3.2 6-Flexagons

The simplest 6-flexagon* is made from a straight strip of 10 equilateral triangles. You may have already constructed it in Chapter 1, but we think it worthwhile to describe it here in terms of instructions that we will use throughout the rest of the chapter. Instructions for the basic construction are these:

1. Prepare the pattern piece, labeling both sides of it *precisely* as shown.
2. Crease all fold lines in *both* directions.
3. Fold *in order* (so that the numbers are no longer visible)
   triangle 1 onto triangle 1,
   triangle 2 onto triangle 2,
   triangle 3 onto triangle 3,
   ⋮
   and finally, triangle ◓ onto triangle ◓ .
4. Glue, or attach with a paper clip, so that ◓ is attached to ◓ .
5. Gently flex and play with your model—decorate the faces with interesting patterns.
6. *Enjoy* your flexagon and show it to your friends!

Here is the pattern piece for the 6-flexagon. Try out the instructions.

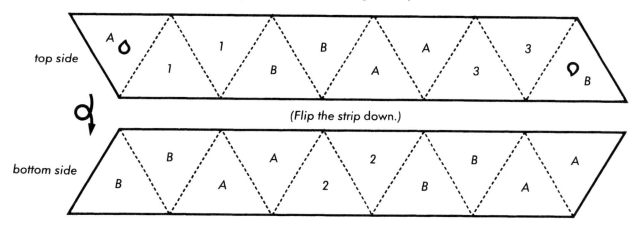

Now for the magic! Gently mountain-fold and valley-fold† the hexagon as shown in the pictures to make a 3-petaled arrangement that will "come apart" at the top and lie flat when the vertices labeled *x, y, z* are brought together *below* the hexagon. Repeat the process. Notice that as you flex the hexagon in this way you see 3 faces, the *A* face, the *B* face, and the 1-2-3 face.

---

*Hexaflexagons were discovered in 1939 by Arthur H. Stone, a Professor Emeritus of mathematics at the University of Rochester. See Martin Gardner's book, *The Scientific American Book of Mathematical Puzzles and Diversions* (New York: Simon and Schuster, 1959), for an interesting account of Stone's discovery and his collaboration with Bryant Tuckerman, then a graduate student, and now a retired research mathematician from IBM (Yorktown Heights, NY), Richard P. Feynman, then a graduate student in physics and later a Nobel Laureate, and John W. Tukey, then a young mathematics instructor and now an Emeritus Professor at Princeton. It is interesting to remark that the diagrams Feynman devised for analyzing 6-flexagons were the forerunners of the *Feynman Diagrams* famous in modern atomic physics.

†A mountain fold is *above* the surrounding terrain; a valley fold is *below* the surrounding terrain.

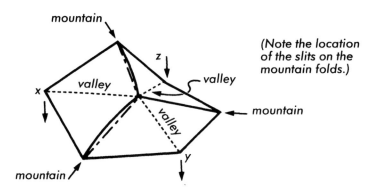

(Note the location of the slits on the mountain folds.)

Although 6-flexagons constructed from the same width of tape will all have the same shape and size, they may differ in the number of hexagonal faces that can be presented as the polygon is flexed. In this sense, the flexagon you have just constructed is the "smallest" hexaflexagon that can be constructed with a *straight* strip of equilateral triangles. Since it has 3 faces, we call it a 3-6-flexagon (it is well known by the name of *tri-hexa-flexagon*).

Play with your flexagon until you become very adept at flexing it. You may want to draw some patterns on its faces. Begin by drawing a design on the two visible faces and then flex it. You will notice a blank face appears and one of the existing faces disappears—and even the face that is still visible may seem changed. You will soon see that although you can draw patterns on only 3 hexagonal faces originally, more than 3 designs will appear, owing to the way the patterns on the triangular portions of the face are moved about when the flexagon is flexed.

A 6-6-flexagon (the *hexa-hexa-flexagon*) may be constructed from a strip of 19 equilateral triangles. Here is the pattern piece. Now it's up to you! Following the basic instructions, make your 6-6-flexagon and then read the rest of this section for suggestions about how to flex it and how to build the even bigger 9-6-flexagons, 12-6-flexagons, and, in general, $3n$-6-flexagons.

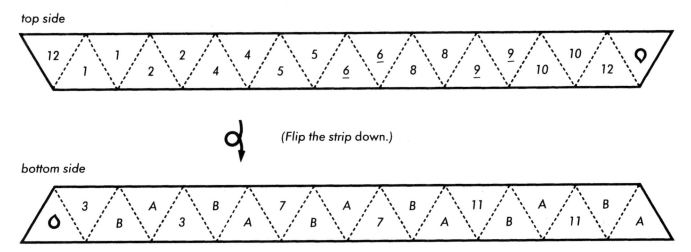

top side

(Flip the strip down.)

bottom side

The 6-6-flexagon is flexed in exactly the same way as the 3-6-flexagon. However, most people have difficulty finding all 6 faces. Bryant Tuckerman invented a procedure for bringing out the 6 faces with the shortest possible flexing sequence. His process, known as the *Tuckerman Traverse*, involves continually flexing at one vertex until the flexagon refuses to open, then moving to an adjacent vertex (either way) and continuing to flex at that vertex until the flexagon refuses to open, ... It is an interesting exercise to record the *sequence of faces* that appears as you perform this flexing algorithm. Try it and compare your results with the diagram in Figure 3.1.

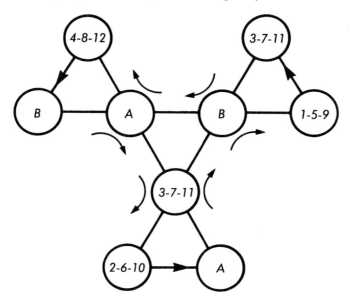

Figure **3.1**

Notice that although you drew 6 patterns on the faces of this polygon, there are many more actual designs (since the triangular parts of the hexagon appear in different orientations as you flex the model). How many different designs do you get from your 6 faces? (Your answer will depend on the symmetry of the patterns you use.)

You may have already noticed a pattern in the *number of faces* on these 6-flexagons. The smaller had 3 faces and the last one had 6 faces.* Do you suppose the next larger one will have 9 faces? If so, how do you construct it? The answer to the first question is yes, and here are some hints about how to go about constructing it.

First of all let us suppose that somehow we remember that it is possible to construct a 3-6-flexagon but that we've forgotten how many triangles we need. We can readily calculate the number of triangles required. We need to have 6×3 triangles available in order to provide the 3 faces. We also need 2 extra triangles that get glued together. Thus this flexagon contains $6 \times 3 + 2$ triangles in all. However, since each triangle on the strip of tape has 2 sides, the number of triangles this model actually requires is only half this number, that is,

$$\frac{6 \times 3 + 2}{2} = 10$$

In exactly the same way, we can reason that for a 6-faced 6-flexagon the number of triangles required is

$$\frac{6 \times 6 + 2}{2} = 19$$

*It is a fact (which we don't prove) that the number of faces that can occur on 6-flexagons constructed with straight strips of equilateral triangles must be a multiple of 3.

We can rewrite each of these expressions (on the left) to see the following pattern:

For the (3×1)-6-flexagon we need a strip of 10 ($= \mathbf{1} \times 9 + 1$) triangles.

For the (3×2)-6-flexagon we need a strip of 19 ($= \mathbf{2} \times 9 + 1$) triangles.

In general:

For the **3n**-6-flexagon we need a strip of ($\mathbf{n} \times 9 + 1$), that is, ($9n + 1$) triangles.

Thus, for example, the 9-6-flexagon (*nona-hexa-flexagon*) requires a strip of $3 \times 9 + 1$ ($= 28$) triangles.

Now that we know the number of triangles required for our 9-6-flexagon, how do we get them folded in the right arrangement? Again we study the two 6-flexagons we've already constructed.

Notice that, in the flattened position of the 3-6-flexagon, the thicknesses of tape on two adjacent triangular sections are 1 and 2, respectively (see Figure 3.2). (Where two triangles are glued together, they behave as 1 thickness of tape.)

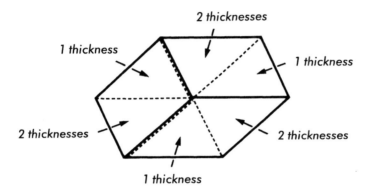

Figure **3.2**   1 + 2 = 3 = the number of faces for a 3-6-flexagon.

What is the situation with the 6-6-flexagon? We observe that in its flattened position immediately after construction (*before* any flexing takes place), the thicknesses of tape on two adjacent triangular sections are 2 and 4. However, when the 6-6-flexagon is flexed, it sometimes has thicknesses of 1 and 5 on adjacent triangular sections. See Figure 3.3.

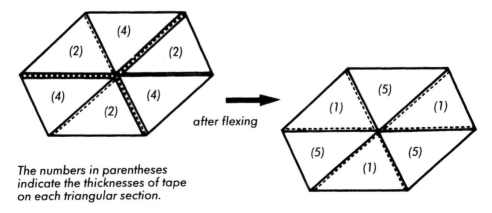

The numbers in parentheses indicate the thicknesses of tape on each triangular section.

Figure **3.3**   2 + 4 = 1 + 5 = 6 = the number of faces for the 6-6-flexagon.

This information contains the secret for constructing the 9-6-flexagon. What we might seek is an arrangement so that the thicknesses on any two adjacent triangular sections of the flattened hexagon sum to 9. One possibility is to use the fact that $4 + 5 = 9$ and try to find out how to fold the strip of equilateral triangles so as to produce adjacent triangles on the finished model having 4 and 5 thicknesses, respectively (see Figure 3.4). But we already know, from our construction of the 6-6-flexagon, how to fold the strip to obtain 4 thicknesses on one of the triangular sections; and, as we've observed, there must exist a way to obtain 5 thicknesses on a triangular section. The idea is to construct the 6-6-flexagon, except that you attach the last two faces together with a paper clip instead of using glue. You can then flex this flexagon until you have thicknesses of 1 and 5 on adjacent triangular sections. At that point you can remove the paper clip and "unwrap" the arrangement to *see* how to fold a triangular section with 5 thicknesses. With a little practice you will be surprised how easily you can guess how to fold the required number of thicknesses for a given triangular section.

In the same way that we figure out from the 6-6-flexagon how to construct the 9-6-flexagon, we can use the 9-6-flexagon to discover how to build the 12-6-flexagon (*dodeca-hexa-flexagon*) with a strip of 37 equilateral triangles. Of course, the process goes on, and you may even find that if you try it you can construct the 15-6-flexagon with a strip of 46 equilateral triangles. In fact, constructing it may be easier for some people than learning how to pronounce its Greek name, which is *penta-cai-deca-hexa-flexagon*! (*Penta-cai-deca* means "5 and 10" or "15" and, of course, "hexa" means "6.")

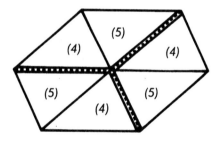

Figure 3.4   Thicknesses for the triangular sections of the 9-6-flexagon.

3.3 8-Flexagons

The 8-flexagon (octa-flexagon), already described in Section 1.7, was discovered independently by many people and is, as we have said, often referred to as a tetraflexagon because the bounding polygon of the original construction is a square. However, we take a different point of view and call this an 8-flexagon because there are 8 hinged triangular sections surrounding the center of the polygon—just as there are 6 hinged triangular sections surrounding the center of the 6-flexagon.

There are many differences (and some similarities) between 6-flexagons and 8-flexagons. The first difference that should concern us is that 6-flexagons are constructed from straight strips of equilateral triangles, whereas 8-flexagons are constructed from straight strips of isosceles right triangles (instructions for folding these appear in Section 1.7).

The simplest 8-flexagon, called a 4-8-flexagon (or *tetra-octa-flexagon*, because it has 4 faces), may be made from a straight strip of 17 isosceles right triangles, as shown in Figure 3.5. (If you have already constructed this model from the instructions in Section 1.7, you may go directly to the flexing instructions). Once the pattern piece is folded, all you need to do to produce your 4-8-flexagon is follow the basic instructions given at the beginning of this chapter.

Just one word of caution before you proceed. This flexagon is much more versatile than the 6-flexagons. It can assume many shapes *other* than the square, and there are at least three different ways it can be flexed. As a result of the 8-flexagon's extraordinary capability to display different faces and shapes to the world, it sometimes gets twisted. This is not serious, and if you are patient, it can always be untwisted. However, if you feel that it is possible that you might suffer from a shortage of patience, we offer this bit of advice. In the beginning just paper-clip the last two faces together, instead of gluing them. This way if you inadvertently get the flexagon in a state that frustrates you, then you can simply remove the paper clip and reassemble it. This procedure also allows you to unwrap the flexagon to find out whether or not you have put patterns on all the faces.

Now get, or construct, your 4-8-flexagon and then return. We'll wait.

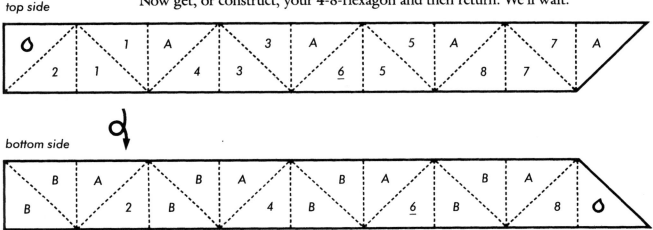

Figure **3.5**

Look at your flexagon. Observe that there are subtle differences between the two visible faces. One side should look like (a) and the other should look like (b).

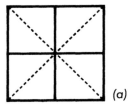 (a)   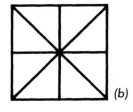 (b)

Surprisingly, this flexagon can *change its shape*. To see how this happens, begin with the (a) side up and execute the moves shown in Figure 3.6. Of course, after you have gone from left to right you will need to reverse the moves to get the flexagon back into its original shape. You may wish to practice these procedures until you have a feel for them. Take your time. Then come back and we'll tell you how to flex your 8-flexagon in ways similar to what you did for the 6-flexagon.

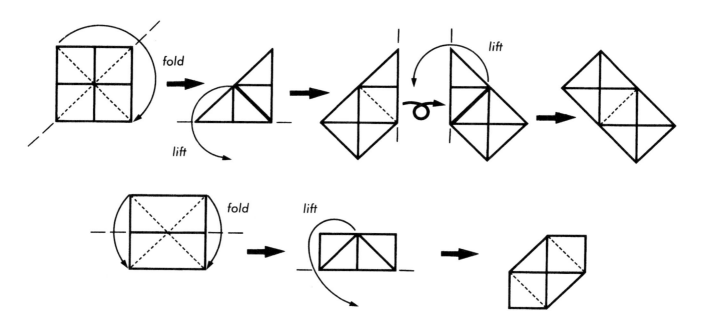

Figure **3.6**

Before we begin flexing the 4-8-flexagon, let us see how we could anticipate how many faces it has. Observe that the thicknesses on adjacent triangular sections are 1 and 3 (see Figure 3.7). Thus we might expect (and it turns out to be true) that this flexagon will have 4 distinct faces. We challenge you to find those 4 faces by repeated flexing.

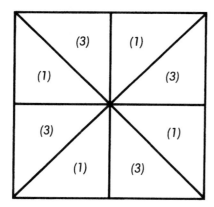

The numbers in parentheses indicate the thicknesses of tape on each triangular section.

Figure **3.7**

We describe first a *straight flex*. Begin with your flexagon in the position shown in Figure 3.8 (if it doesn't look exactly like this, turn it over). Then make mountain folds along the dotted lines, and bring the 4 vertices labeled *A*, *B*, *C*, and *D* together below the flexagon. The top of the flexagon (surrounded by the vertices *E*, *F*, *G*, and *H*) will then "come apart" and the flexagon will lie flat again in the shape of a square. To repeat this straight flex, as we call it, you must turn the flexagon over. Practice this a few times and draw patterns on all the faces you can find.

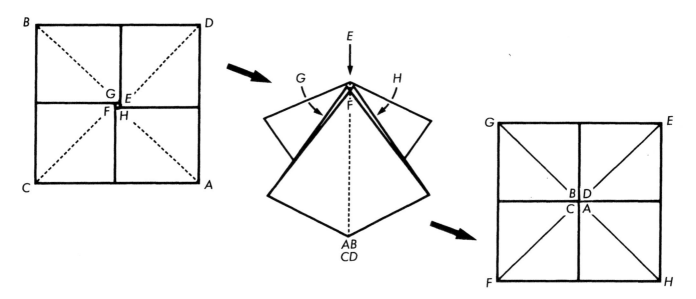

Figure **3.8**   Executing the *straight flex*.

You are now ready for the more complicated *pass-through flex*. To do this, begin with the other side of the flexagon up (as shown in Figure 3.9) and make mountain folds along the diagonal lines so that you obtain a 4-petaled arrangement. Then pull two opposite petals down. You will then have a square platform above the two petals you pulled down. Fold the sides of the platform down and let the flexagon open at the top as shown in Figure 3.9. Be patient, keep calm, stay happy! It will work! (In the very unlikely event that you don't get it right the first time, try again—the very worst that can happen is that you will have to take your flexagon apart, reassemble it, and then try again.)

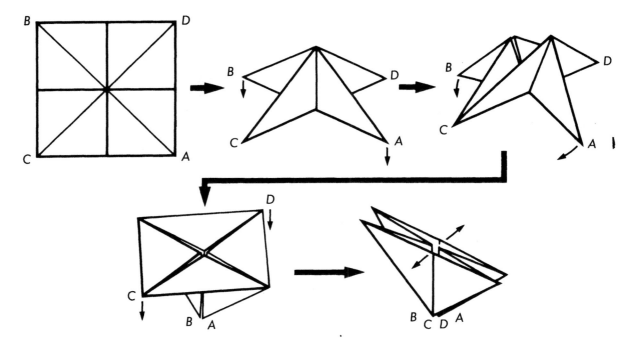

Figure **3.9**   Executing the *pass-through flex*.

Practice flexing with the straight and pass-through flexes. How many faces can you find? (*Answer:* Only 3.) However, there is yet another way to flex. We call it the *reverse pass-through flex,* and we've saved it for last because it really is a little tricky. To perform this flex you do all the steps of the pass-through flex, but just at the point where you would open the flexagon, *stop!* At this point your flexagon looks like 4 petals that are formed by 4 mountain folds and 4 valley folds. Now you complete the reserve pass-through flex by *reversing* the mountain and valley folds (as shown in Figure 3.10) and then opening the flexagon at the top as before. *Now* you will see the fourth face.

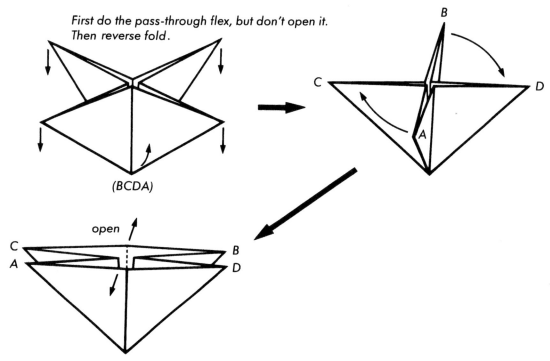

Figure **3.10**   Executing the *reverse pass-through flex.*

After you have become familiar with your 4-8-flexagon you will be ready for a real treat—the 8-8-flexagon (*octa-octa-flexagon*). This model is no more complicated to construct than the 4-8-flexagon, and it is flexed in exactly the same way.

We now have a special challenge for you. Recall that we know how to bring out all the faces on a 6-flexagon with a simple algorithm (the Tuckerman Traverse). We therefore felt sure that there must be a fairly simple algorithm for bringing out all the faces of an 8-flexagon. In fact, such an algorithm was shown to us by Jennifer Hooper (the Jennifer Pedersen of Chapter 6), but we should warn you that although each face appears on the top *or* the bottom in the course of executing this algorithmic sequence of flexes, it is *not* true that each face appears on the top, nor is it true that each face appears on the bottom. Can you find Jennifer's algorithm? We'll give you a hint. Her algorithm involves only the straight flex and the reverse pass-through flex. But this is not surprising, since, as you may verify, the pass-through flex has the same effect as the following sequence of three flexes:

reverse pass-through flex
straight flex
reverse pass-through flex

Figure 3.11 gives the pattern piece for the 8-8-flexagon.

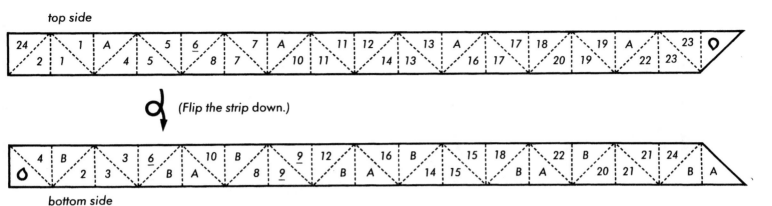

Figure **3.11**

As you construct your 8-8-flexagon, you may notice that the thicknesses on adjacent triangular sections are 3 and 5 (which accounts for the 8 faces). Notice, however, as you flex your 8-8-flexagon, that it is possible to produce an arrangement where the thicknesses on adjacent triangular sections will be 1 and 7 (see Figure 3.12). This is marvelous! It means that, just as we could figure out how to build 6-flexagons with $3n$ faces from 6-flexagons with fewer faces, we can figure out how to build 8-flexagons with $4n$ faces from 8-flexagons with fewer faces.

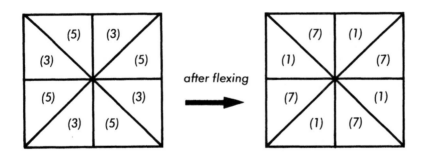

Figure **3.12**

We now close this chapter and leave the exploration of these ideas to you. We feel certain some of you will want to take a strip of

$$\frac{12 \times 8 + 2}{2} = 49$$

isosceles right triangles and build the 12-8-flexagon (*dodeca-octa-flexagon*), which has 5 and 7 thicknesses on adjacent triangular sections. You may even want to take a strip of

$$\frac{16 \times 8 + 2}{2} = 65$$

isosceles right triangles and construct the 16-8-flexagon. We could even believe that some people might want to make this giant flexagon just so that, when there is a lull in the conversation, they can talk about their *hexa-cai-deca-octa-flexagon!*

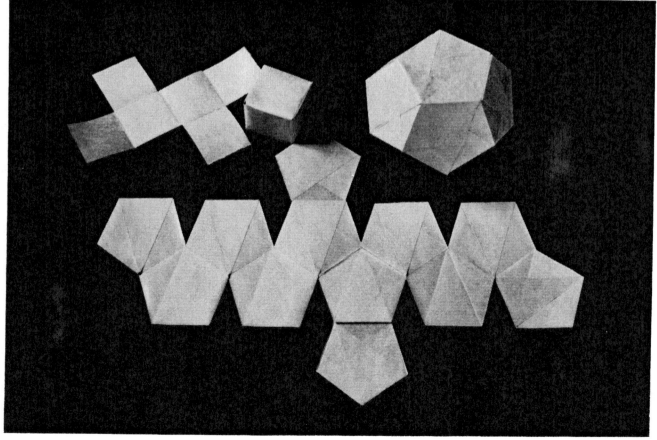

4 Introduction to Polyhedra

4.1 An Intuitive Approach to Polyhedra

In Chapter 1 we looked at polygons in the plane and, in particular, we studied how to construct regular convex polygons. A natural extension of this idea in 3 dimensions is to study how to construct *polyhedra*, which are, in an obvious intuitive way, the 3-dimensional analogues of polygons. For example, just as a connected polygon divides the plane into two regions (the inside and the outside), a connected polyhedron divides space into two regions (again, the inside and the outside). A polygon consists of straight (uncurved) sides, that is, parts of lines, whereas a polyhedron consists of flat (uncurved) faces, that is, parts of planes. A rectangular box is an example of a polyhedron, but a cylindrical can of soup is not, because its entire boundary does not consist of flat faces.

The formal study of polyhedra is very rich and intellectually rewarding, but we will restrain ourselves and postpone our general discussion of the mathematics until the last chapter. Here, and in the next chapters, we study polyhedra from the practical, constructive point of view. We think this is the appropriate order of events. You will discover that making the models we describe is a vivid and educational experience. Sometimes it will seem almost magical when the faces finally all fit together—and sometimes the final shape obtained is surprisingly beautiful. Of course, it can be exasperating if the pieces don't fit together correctly, but we've tried to spare you this unpleasant experience by including in our instructions more information than is actually needed to construct the polyhedra, some of it of a very practical, nonmathematical nature.

Nevertheless, most people find constructing polyhedra so exciting and absorbing that it would be quite pointless to try to do *anything* else at the same time. Teachers, especially, should be aware of this and should let their students enjoy the experience of constructing the polyhedra without other distractions. Don't be too impatient, wait, and discuss the mathematical properties of the models *after* some models have been constructed.

In the next section we describe a classical method for constructing polyhedra from what are called *net diagrams*, and we make some suggestions for how you can use what you learned in Chapter 1 to produce the nets. When you have constructed the models, you will be prepared to read Section 4.3, where we ask (and answer) the questions, What is a polyhedron? and What is a *regular* convex polyhedron?

But let us get on with constructing some polyhedra so that you can begin to ask some questions of your own and be able to appreciate the definitions we give in Section 4.3.

4.2 Constructing Polyhedra from Nets

Required Materials

☐ Small pieces of cardboard (from shoe boxes, for example)
☐ 8½ × 11 in. (or bigger) sheets of heavy construction paper (colored paper is nice)
☐ Pen or pencil
☐ White glue
☐ Scissors

Optional Materials

☐ Gummed mailing tape about 2 in. wide (the heaviest quality you can find)
☐ Sponge
☐ Bowl of water
☐ Hand towel (or rag)
☐ Books
☐ Colored paper, acrylic paint, or glitter

The idea, which is especially suitable for young children (provided that an adult prepares the pattern pieces!), is simple. Think of taking a polyhedron, such as a cube, that is made from paper and slitting apart some of the edges so that all of the faces of the polyhedron lie flat and the whole thing still remains in one piece. The resulting configuration (of which, for any given polyhedron, there are many possibilities) is called a *net* for the polyhedron. Now, since our object is really to go the other way, that is, to *construct* the polyhedron from a net, it is necessary to add some tabs to the net so that we can glue the edges together. It is an interesting and useful fact that for any net of a polyhedron, if tabs are attached to alternate sides of the boundary of the net, then it will always be possible to assemble the polyhedron by using those tabs to join appropriate faces. Notice that we said *appropriate* faces. We were very careful about this because, as you may easily verify—in the case of the triangular dipyramid, for example—it is sometimes possible to join the faces of the net by means of the tabs in such a way that you don't get the desired polyhedron—or, in fact, any polyhedron at all! Where we think it would be helpful, we have added arrows to the net to indicate that the two sides at the beginning and end of the arrow should be joined to form an edge of the polyhedron.

Figure 4.1 shows nets, with the tabs attached, for a special class of polyhedra called *convex deltahedra* (we explain the word *convex* in the next section; a *deltahedron* is a polyhedron, all of whose faces are triangles). An illustration of the completed model (and its name) is shown next to each net so that you will have an idea of how the completed polyhedron should look.

General Instructions (Traditional)

1. Make a cardboard pattern in the shape of an equilateral triangle. See Section 1.3 for a method of folding an equilateral triangle if necessary.
2. Decide which polyhedron you want to make.

3. Use your pattern piece to duplicate the required net on a piece of construction paper. Press firmly with your pen or pencil so that it will be easy to fold along the edge lines. The shape of the tabs is not critical, as long as they do not extend beyond the sides of the triangle you are using for a pattern.

4. Cut out the pattern piece and carefully valley-fold along each of the pen (or pencil) lines.

5. Glue the tabs in place. Notice that, because you made *valley* fold lines on the pen (or pencil) lines, the lines will be on the inside of the model. This will give a cleaner looking model.

6. Cut out colored triangles and glue them on the faces, paint the faces with acrylic paint, or spread glue on the faces and sprinkle glitter on them. In a word, be creative! (optional)

General Instructions (Alternative)

1. Select the polyhedron you want to make.

2. Count the number of faces and tabs.

3. Fold a strip of heavy gummed tape *UDUD...* until you have about twice as many triangles as there are faces and tabs on the model you want to make. (Keep any extra triangles for the next model.)

4. Cut and glue portions of the *UD* tape together to make a duplicate of the desired pattern piece. Do this so that the gummed side is on the inside of the finished model. Here it is particularly easy to make tabs in the shape of equilateral triangles; for ease of construction, mark them so that you will know they are tabs and not faces.

5. Place heavy books on the pattern piece and tabs while they dry. This keeps the faces flat.

6. Fold and crease all the edges so that the edges will be sharp.

7. Glue the tabs in place (paying attention to the arrows, if there are any on the net you chose).

8. Color the faces. (optional)

Notice that the triangular tabs add strength to the model and, since they cover the entire face, their function is undetectable. Notice, also, that with this method all of the gluing is done on the *outside* of the model. We think this is a distinct improvement over those instructions that tell you to glue the tabs *inside* the polyhedron. (We always find that the last tab is then almost impossible to do neatly, if it can be done at all.)

Now try it!

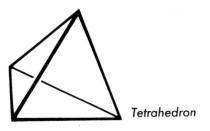

Tetrahedron

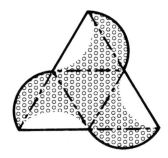

Figure **4.1**   Convex Deltahedra (Based on patterns from *Geometric Playthings* by Jean J. and Kent A. Pedersen.)

Figure **4.1** cont.

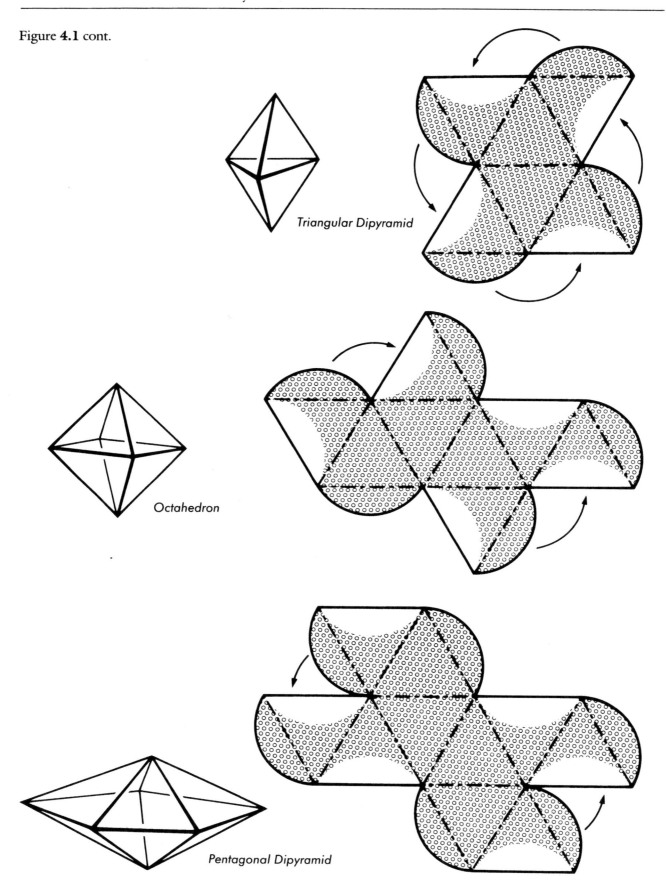

Triangular Dipyramid

Octahedron

Pentagonal Dipyramid

Figure **4.1** cont.

Dodecadeltahedron

Tetracaidecadeltahedron

Figure **4.1** cont.

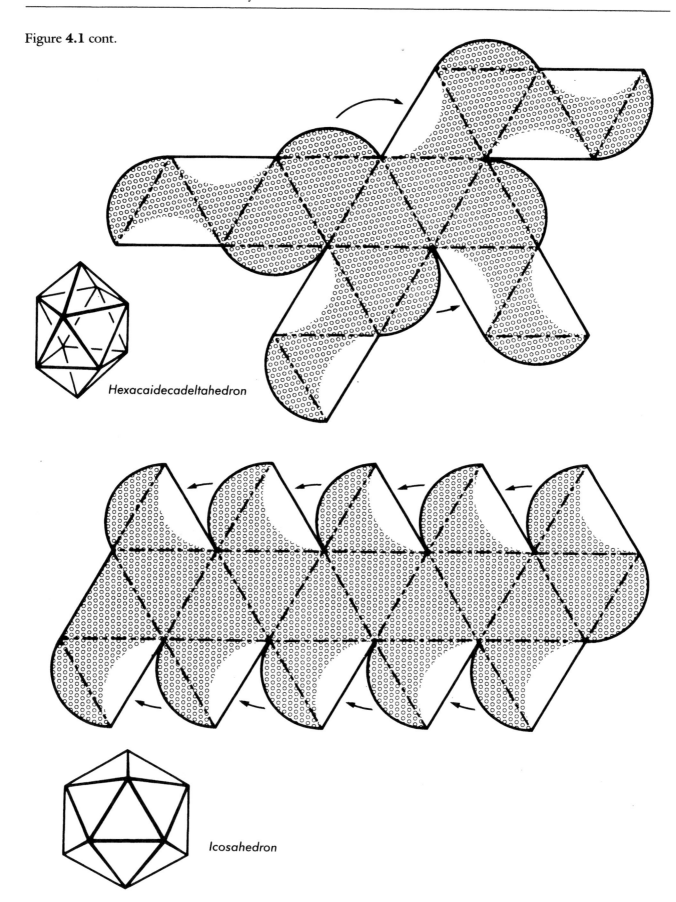

Hexacaidecadeltahedron

Icosahedron

Notice that the idea of constructing polyhedra from nets works just as well if the faces are regular squares or pentagons. So we now give you the nets for the hexahedron, or cube, and the dodecahedron and leave the necessary modification of the Traditional General Instructions to you (see Figure 4.2). We give one minor word of advice. The tabs are not shown on these nets; but, in the case of the cube, you proceed by making one square cardboard pattern piece (see Section 1.7 if you've forgotten how to fold a square) and drawing a net. Then add the tabs to alternate sides of the boundary of the net to produce a pattern for the cube. You may wish to use tabs that are themselves square, since they will then reinforce the construction and leave no trace of their function.

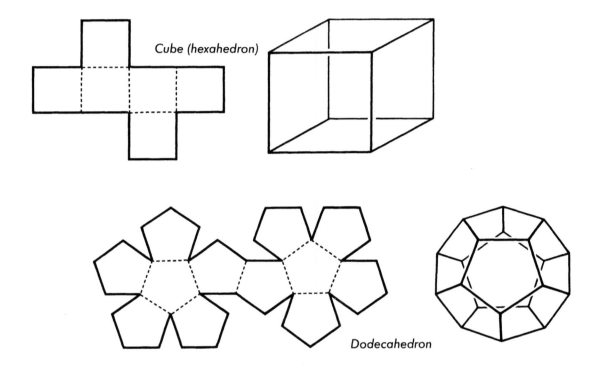

Cube (hexahedron)

Dodecahedron

Figure **4.2**

In a similar way, if you begin with one cardboard pentagon (you can produce the pattern for it by the method of Section 1.4), you can draw the net for the dodecahedron and add the tabs to alternate sides of the boundary to form the desired pattern piece. Here, however, it is not possible for all the tabs to be in the shape of a regular pentagon. (Do you see why?) However, if the size of the pentagon you used to make your net is the same as that of the pentagon formed by folding along short lines of the U^2D^2 tape, then you can use more of this same U^2D^2 tape to cut out tabs to glue onto the net diagram (using a sponge and water). Simply cut along every other short line, as shown here, to get the tabs.

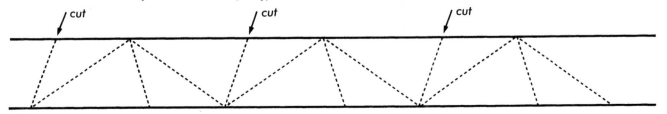

Alternative Construction

It is also possible to construct a net from a single U^2D^2 strip of tape. First think of a *short section* of the U^2D^2 tape

as this 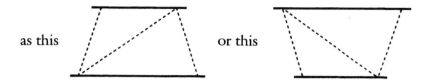 or this

Now take a strip of tape containing 30 short sections and fold it along certain short lines, gluing all the overlapping portions, so that the result looks exactly like this (a small sponge and a bowl of water are handy—with a hand towel or rag to wipe up the excess water):

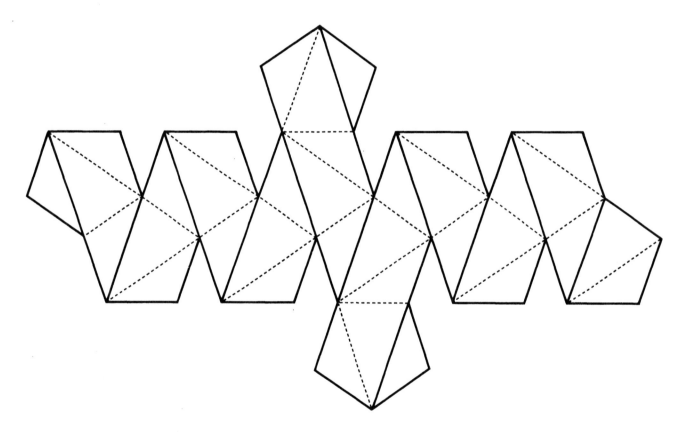

The only tricky part here is to realize that the pentagons at the top and bottom of this figure are formed from *five* short sections (zigzagging back and forth to form a very strong pentagon). Once the net is prepared, tabs can be made, as before, by cutting the U^2D^2 tape on every other short line.

Practical Hints

1. Be certain to glue all of the overlapping portions.
2. Let the net with its tabs attached dry underneath some heavy books so that the faces will be as flat as possible.

3. Crease all of the edge lines firmly so that the edges will be sharp and straight.
4. Where tabs are attached, crease the edge firmly before gluing the other half of the tab to the model.
5. Be patient; don't try to go too fast. Wait until each tab is stable before gluing the next one.

Now that you have constructed some polyhedra, you should get to know them. Play with them, admire them, and notice how many faces they have, what they look like from various directions, and how they *feel*. Do you notice anything about them when you close your eyes and hold them that you didn't notice by simply looking at them? Experiment by holding them in front of a light and looking at the shadows they cast. Try classifying them according to certain properties.

4.3   *What Is a Regular Polyhedron?*

Before we attempt to answer this question we should be more precise as to what we mean by a polyhedron.

Let us begin by looking again at polygons. We distinguish between a *polygon* (or polygonal path) and a *polygonal region*. A polygon consists of edges hinged together at vertices and does not contain its interior (recall that it was the top edge of the tape in Chapter 1 that formed the polygon). Similarly, we say that a polyhedron consists of faces hinged together at edges, and it does not contain its interior. Thus, strictly speaking, a polyhedron is a *surface* and not a *solid*, although it is sometimes loosely referred to as a solid (for example, the Platonic Solids, which we will discuss shortly).

Now, just as each vertex of a polygon is an endpoint of exactly two edges, so is each edge of a polyhedron the side of exactly two faces. Thus a polyhedron should be regarded as a collection of faces, each of which is a polygonal region; and two intersecting faces intersect precisely in a common side of each, or in just one vertex that is a common vertex of each face. For examples, look at the figures next to the net diagrams in Figure 4.1.

A polygon is *connected*, meaning that it is all in one piece; the polyhedra we consider are also connected in this sense. Further, just as we lay particular emphasis on *convex* polygons, so too we confine our attention here to *convex* polyhedra. We define a polyhedron to be *convex* if, given any two points P and Q of the region bounded by the polyhedron, then every point of the straight line segment PQ belongs to that region. Alternatively, we may say that if P and Q are any two points of the polyhedron, the straight-line segment PQ consists of points of the polyhedron or its interior.

So in this section we will be discussing convex connected polyhedra as here defined. We emphasize that this is a very restrictive definition. In particular, the restriction to convex polyhedra is far more significant than the corresponding restriction to convex polygons. Without this restriction, it is not even meaningful to talk of "the region bounded by the polyhedron." You may think of the polyhedra we wish to discuss as being obtained from a spherical surface made of some plastic material by "pushing and pulling" it around until it consists of flat polygonal faces as described.

It is customary to name polyhedra in a manner similar to the way we named polygons—that is, just as we incorporated the number of sides of a polygon into its name (thus a *penta*gon has 5 sides), we now incorporate the number of faces a polyhedron possesses into its name.* We next list the names of some of the better-known polyhedra. You will note that the part preceding *hedron* designates the number of faces (*poly* means "many"). In the case of a convex polyhedron, each of these faces can be used as a base when a model of it is set on a table. This explains the use of the generic term *hedron*, which is the Greek word meaning "base" or "seat."

Some Names

A polyhedron with:	is called a:
4 faces	tetrahedron
5 faces	pentahedron
6 faces	hexahedron
7 faces	heptahedron
8 faces	octahedron
10 faces	decahedron
12 faces	dodecahedron
14 faces	tetracaidecahedron (*cai* means "and")
15 faces	pentacaidecahedron
16 faces	hexacaidecahedron
20 faces	icosahedron

In Chapter 1 we defined a *regular* convex polygon (with 3 or more sides) to be a polygon with all sides equal and all angles equal (you may recall that only in the case of the triangle does each of these conditions imply the other). So now we ask: **What would be an appropriate analogous requirement for a *regular* polyhedron?**

We reason that, since the faces of polyhedra are analogous to the sides of polygons, it should make sense to require that every face on a regular polyhedron should be the same regular polygon. We can readily agree, however, that this wouldn't be a strong enough requirement. For example, observe that the polyhedra that were constructed entirely of equilateral triangles are not all really regular. Thus, if you view the triangular dipyramid from the top vertex you see 3 triangles meeting, but if you view it from a side vertex you see 4 triangles meeting. A regular polyhedron should surely look the same when viewed from any vertex (or from any edge or from any face). So it would be reasonable to require that the arrangement of the polygonal faces on a regular polyhedron should be exactly the same at every vertex. Let us impose this extra restriction and then see, first of all, if any polyhedra exist that satisfy our requirements and, second, if the resulting polyhedra deserve to be called "regular."

Following this idea we let p stand for the number of sides of each regular face. Then we know, from Section 2.1, that the interior angle of the regular p-gon, measured in degrees, is

$$\frac{(p-2)180}{p}$$

---

*It is common, in ordinary language, to refer to the *sides* of a cube; thus "a cube has 6 sides." We strongly advise you to avoid this very misleading use of the word *side* and to use instead the mathematician's word *face*; thus "a cube has 6 faces." We use the word *side* exclusively to refer to a line segment forming part of the boundary of a face. Thus, for us, a cube has 6 faces, and each face has 4 sides.

Now if we consider the arrangement of q of these regular p-gons about a single vertex, we see that the sum of the face angles about that vertex is

$$q\left(\frac{(p\text{-}2)180}{p}\right)$$

Notice that $p \geqslant 3$. However, we also have $q \geqslant 3$, because at least 3 faces must come together at each vertex.

The concept of convexity now becomes important. It is a fact, first remarked by Euclid (which we illustrate but do not prove), that the sum of the face angles about any vertex on a *convex* polyhedron must be less than 360°. All the models you have constructed in this section illustrate this fact. However, we see from Figure 4.3(b) that the converse is not true, that is, the sum of the face angles at every vertex may be less than 360° without the polyhedron being convex.

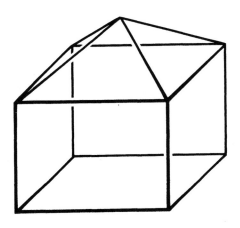
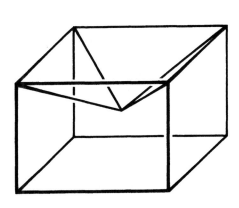

Figure **4.3**   (a) Convex polyhedron   (b) Nonconvex polyhedron with the same face angles at each vertex as its convex cousin on the left.

So now let us look for values of p and q, with $p \geqslant 3$ and $q \geqslant 3$, such that

$$q\left(\frac{(p-2)\,180}{p}\right) < 360$$

or

$$pq - 2q < 2p$$

after dividing by 180 and multiplying by p. Straightforward algebra then gives the following sequence of equivalent inequalities:

$$pq - 2p - 2q < 0$$
$$pq - 2p - 2q + 4 < 4$$
$$p(q-2) - 2(q-2) < 4$$
$$(p-2)\,(q-2) < 4$$

Notice that this last expression is symmetric in p and q—and thus if you find any values for p and q that satisfy this inequality, you can find the "complementary" solution by exchanging the values for p and q. Notice also that since $p \geqslant 3$ and $q \geqslant 3$, it follows that $p - 2 \geq 1$ and $q - 2 \geq 1$. Thus if $(p-2)(q-2) < 4$, then $(p-2)(q-2)$ must be 1, 2, or 3. When the product $(p-2)(q-2)$ is 1 we must have

$$p - 2 = 1 \text{ and } q - 2 = 1$$

so that we obtain $p = 3$ and $q = 3$.

If $(p-2)(q-2) = 2$, then either

$$p - 2 = 2 \text{ and } q - 2 = 1$$

or

$$p - 2 = 1 \text{ and } q - 2 = 2$$

Thus, in the first instance, $p = 4$ and $q = 3$; in the second instance, $p = 3$ and $q = 4$.

Finally, if $(p-2)(q-2) = 3$, then either

$$p - 2 = 3 \text{ and } q - 2 = 1$$

or

$$p - 2 = 1 \text{ and } q - 2 = 3$$

The first equations give $p = 5$ and $q = 3$; the second equations yield $p = 3$ and $q = 5$.

Thus we see that the only possible solutions to our inequalities $(p-2)(q-2) < 4$, $p, q \geq 3$ are

$$p = 3, q = 3$$
$$p = 3, q = 4$$
$$p = 4, q = 3$$
$$p = 3, q = 5$$
$$p = 5, q = 3$$

In the form of a table, these give

(p, q)
$(3, 3)$
$(3, 4)$
$(4, 3)$
$(3, 5)$
$(5, 3)$

In fact, each solution corresponds to an actual polyhedron—and you have already been given instructions for constructing them. Each value of (p, q) determines exactly one polyhedron. The polyhedra in this very special set are known as the *Platonic Solids* (see Figure 4.4). We remind you that, as described, they are not really solid because they do not contain their interiors. But we also note, with some pleasure, that they satisfy our criteria for regularity; indeed, nobody could conceivably dispute the claim that they are regular! Thus we adopt our provisional definition of regularity, and we are in the happy position of knowing all the regular polyhedra—they are just the five Platonic Solids.

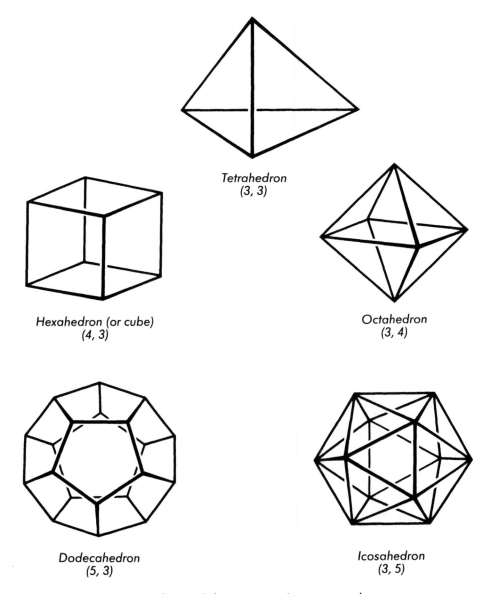

Tetrahedron
(3, 3)

Hexahedron (or cube)
(4, 3)

Octahedron
(3, 4)

Dodecahedron
(5, 3)

Icosahedron
(3, 5)

*The notation (p, q) means that each face is a regular p-gon and
q faces come together at each vertex.*

Figure **4.4**   The Platonic Solids.

REFERENCES

Benson, Jean, Peter Hilton, and Jean Pedersen, *College Preparatory Mathematics*. New Almaden, CA: Pedersen Publishing, 1993. Many of the ideas in this chapter first appeared in Chapter 4 of an early version of this book.

Billstein, Rick, et al. *A Problem Solving Approach to Mathematics for Elementary Teachers*, 3rd ed. Menlo Park, CA: The Benjamin/Cummings Publishing Co., 1987.

Pedersen, Jean J., and Kent A. Pedersen. *Geometric Playthings*. Palo Alto, CA: Dale Seymour Publications, 1986.

5

Constructing

Dipyramids from a

Single Straight Strip

Required Materials

- ☐ About 5 ft of 2-in. gummed mailing tape (or a longer length of wider tape if you want a larger model). The glue on the tape should be the type that needs to be moistened to become sticky. Don't try to use tape that is sticky to the touch when it is dry—unless you want an exercise in frustration.

Optional Materials

- ☐ Scissors
- ☐ Sponge (or washcloth)
- ☐ Shallow bowl
- ☐ Water
- ☐ Hand towel (or rag)
- ☐ Some books
- ☐ Colored paper of your choosing. Construction paper works well, but it must be cut into strips and glued together to get long enough pieces. Gift wrapping or butcher paper that comes in rolls is particularly easy to use for these models.

5.1 Preparing the Pattern Piece

Begin by folding the gummed mailing tape precisely as shown in Section 1.3. Do this with the gummed side *up*, so you can see your fold lines better. Continue folding until you have 40 or more triangles. Cut or carefully tear the tape on the last fold line. Then cut off, or tear off, a strip containing 31 triangles counting from this end (that is, *not* from the end from which you began folding).

Next place your strip of 31 equilateral triangles so that one end appears as shown in Figure 5.1 with the *gummed side down*. Mark the first and eighth triangles *exactly* as shown (note the orientation of each of the letters within their respective triangles).

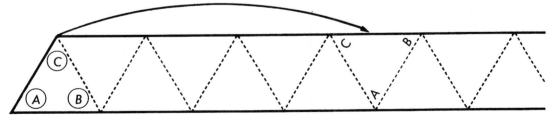

Figure **5.1**   Left-hand end of pattern piece.

5.2 Assembling the Model

Begin by placing the first triangle *over* the eighth triangle so that the corner labeled Ⓐ is over the corner labeled *A*, Ⓑ is over *B*, and Ⓒ is over *C*. Hold these two triangles together, in that position, and observe that you have the beginning of a double pyramid for which there will be five triangles above and five triangles below the horizontal plane of symmetry, as shown here. You may wish to secure the overlapping triangles by moistening just a small part of the center of the gummed triangle.

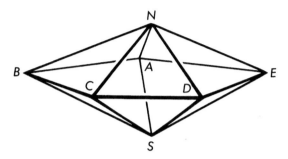

Figure **5.2**   Completed pentagonal dipyramid.

Now you can hold the model up and let the long strip of triangles fall around this frame. If the strip is folded well, the remaining triangles will fall into place. When you get to the last triangle, there will be a crossing of the strip that the last triangle can tuck into, and *your pentagonal dipyramid is complete!* See Figure 5.2.

If you have trouble because the strip doesn't seem to fall into place, there are two frequent explanations. The first (and more likely) problem is caused by not folding the crease lines firmly enough. This situation is easily remedied by refolding each crease line with more conviction, and it is sometimes helpful to fold each crease line in both directions (so that each fold line is scored as both a *mountain* and a *valley* fold). The second common difficulty occurs when the tape seems too short to reach around the model and tuck in. This problem can be remedied by trimming off a tiny amount from each edge of the tape (see Section 6.3 for an explanation of this phenomenon).

5.3 Variations

If you want to make a more attractive model, you may glue the strip of triangles onto a piece of colored paper. To do this, first prepare the piece of paper on which you plan to glue the *prefolded* strip. Make certain it is long enough and that it all lies on a flat surface.

Place a sponge (or washcloth) in a bowl. Add water to the bowl so that the top of the sponge is very moist (squishy).* Moisten one end of the strip of triangles by pressing it onto the sponge; then, holding that end (yes, it's messy!), pull the rest of the strip across the sponge. Make certain the entire strip gets wet and then place it on the colored paper. Use a hand towel (or rag) to wipe up the excess moisture and to smooth the tape into contact with the colored paper.

*Perhaps some very old clothes should be included among the optional materials, if not among those required!

Put some books on top of the tape so that it will dry flat. When the tape is dry, cut out the pattern piece, trimming off a small amount of the gummed tape (about ⅟₁₆ to ⅛ of an inch) from the edge as you do so (this serves to make the model look neater and, more importantly, allows for the increased thickness produced by gluing the strip to another piece of paper). Refold the piece so that the raised ridges are on the colored side of the paper.

You can now construct the polyhedron exactly as before—except that now you probably won't need to label any of the triangles. If you find it difficult to make the last triangle tuck in because it won't reach, you should trim off a little more from each edge.

Next, observe from Figure 5.2 that the completed pentagonal dipyramid has a well-defined equatorial pentagon *ABCDE* going around its middle. Each edge of this equatorial pentagon is incident with exactly two triangular faces (for instance, *AB* is incident with face *NAB* and face *SAB*), and each triangular face is incident with exactly one edge of the equatorial pentagon. The two faces sharing an equatorial edge are said to be *associated*. If the faces are labeled at random with the digits 0, 1, 2, 3, 4, 5, 6, 7, 8, 9, then we may use the pentagonal dipyramid as a die. When we throw the pyramid, it comes to rest on one face, and we declare the number on the *associated* face to be the result of the throw. In this way this die becomes a device for generating random numbers in base 10. There is a slight bias due to the fact that all but one of the faces of the die are covered by exactly three triangles from our original strip, whereas one face is covered by four triangles. However, this bias can be virtually eliminated by trimming off half of the first and last triangles (so that the pattern piece becomes a rectangle) and pretending the whole triangle is there when you assemble it.

You may wish to figure out how to make the analogous construction of a *triangular* dipyramid from a single strip of 19 equilateral triangles. If you want a fair random-number generator for the numbers 0, 1, 2, 3, 4, 5, trim off half of the first and last triangles. Knowing that the finished model should appear as shown in Figure 5.3 and that you should begin by forming the *top* three faces with one end of the strip should get you off to a good start.

Figure **5.3**   Completed triangular dipyramid.

You may discover that there are ways of constructing these dipyramids with fewer than the number of triangles specified, but the real question is, Will they be balanced, in the sense that every face is covered by the same number of triangles? It is not a difficult question to answer—and it is therefore left to the reader as an exercise.

The authors wish to thank FILM IDEAS for granting permission to use ideas in this chapter that originally appeared in the booklet accompanying their video tape "Mathematics for Tomorrow's World."

6 Constructing Jennifer's Puzzle

Required Materials

- ☐ Strips of paper (preferably of different colors)
- ☐ Heavy paper, such as lightweight cardboard
- ☐ Paper clips

Optional Materials

- ☐ Ruler
- ☐ Compass

6.1 Facts of Life

In many instances involving the *use* of geometry in the real world, we need to make adjustments that take into account the *realities of life*. For example, paper comes in various thicknesses (which are never zero!) and the interior of every container must be *larger* than what it contains. These and other very elementary facts of reality affect *how* we are able to take practical advantage of the theorems obtained from our study of the geometry of idealized lines, planes, and solids.

As the section title implies, we concern ourselves here with the details of *practical construction*, in this case of a particular set of nested polyhedra. Namely, we construct an octahedron and 4 tetrahedra that fit inside a larger tetrahedron that, in turn, fits inside a cube. As you will see—assuming that you become actively involved in carrying out these constructions—overcoming the difficulties encountered in using, in a real-life situation, a theory that is perfect in principle is very much a skill of the eyes and hands as well as of the mind.

What follows is simply a description of the construction, along with some hints about how to *solve* Jennifer's puzzle; we do not discuss the many important mathematical insights to be gained from the solution. After completing the construction and assembling your puzzle, you may wish to consult Chapter 12 for some of the mathematical consequences of solving the puzzle. The details have, in fact, already appeared in a Filipino journal (see the references at the end of this chapter), which may not be readily accessible to you.

6.2 Description of the Puzzle

The puzzle consists of 17 strips of paper and an instruction sheet. Figure 6.1 and the instruction sheet from Jennifer's original puzzle identify the 17 puzzle pieces and, simultaneously, tell what is required in the solution.

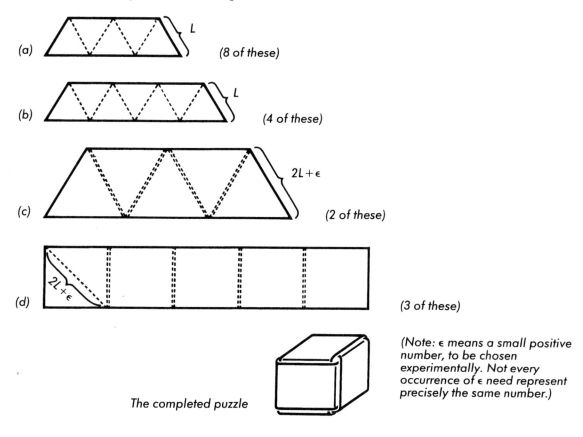

Figure **6.1**

The completed puzzle

(a)   L   (8 of these)

(b)   L   (4 of these)

(c)   2L + ε   (2 of these)

(d)   2L + ε   (3 of these)

(Note: ε means a small positive number, to be chosen experimentally. Not every occurrence of ε need represent precisely the same number.)

INSTRUCTIONS FOR JENNIFER'S PUZZLE: TRY IT!

1. You get all the little strips of 5 triangles each (there should be 8) and braid them into 4 tetrahedra.
2. Then you get the 4 strips of 7 triangles each and braid an octahedron (that is, an 8-faced polyhedron).
3. Now you take the 2 big strips of 5 triangles each and braid a large tetrahedron as before, but in this one you put the 4 little tetrahedra and the octahedron.
4. Finally, take the 3 strips of 5 squares each and braid a cube, into which you put the large tetrahedron.

GOOD LUCK!

Jennifer Pedersen
9th Grade Geometry Project
Castillero Junior High School
San Jose, California

6.3 *How to Make the Puzzle Pieces*

The choice of material for the strips shown in Figure 6.1(a) and (b) is not of great importance, as long as the material has enough bulk and crispness to hold a good fold. Of course, the puzzle will be more interesting visually if you use paper of different colors for different strips; that is, use a different color for each of the *two* strips that form a small tetrahedron, and for each of the *four* strips that form the small octahedron.

The pattern pieces should be scored so that all the dotted lines are valley folds. One way to do this is to start with a strip *longer* than you really need and draw an angle of 60°, as shown:

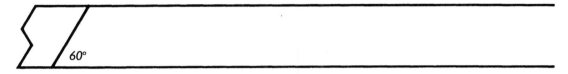

Then fold the paper down:

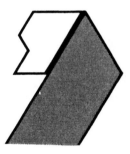

Then unfold:

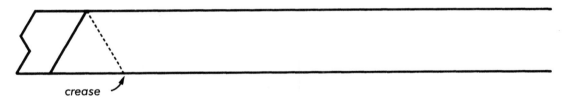

Notice that along the fold line there is a small width (crease) of paper—this crease is usually easier to see on the underside of the paper. The width of the crease will depend on the thickness of the paper used; for this reason it is very important that the next fold line be made precisely as shown:

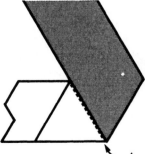

Avoid covering any part of the crease produced by the previous fold line.

Continue folding this way until you have the required number of triangles for steps 1 and 2 of the instructions for Jennifer's puzzle.

For the large tetrahedron and the cube it is necessary to use heavier paper, such as lightweight cardboard. Here, again, this poses problems along the fold lines. The thicker the paper, the more pronounced these difficulties become. What is required is mostly an awareness of the problem so that when you draw the pattern pieces you leave room for the hinge.

For the large tetrahedron, begin by determining, experimentally, the length $2L + \epsilon$, as shown in Figure 6.1(c). This is easily done by placing the completed octahedron and its 4 tetrahedra together, as shown in Figure 6.2, and *measuring* the length that will be required in order for them to *fit inside* the bigger tetrahedron. Then, using a ruler and compass or some other method (for example, paper-folding), construct, on a lightweight piece of paper, an equilateral triangle of the appropriate edge length. This *pattern triangle* may then be cut out and used to draw the big strip of 5 equilateral triangles, as shown in Figure 6.1(c). The width of the space between successive triangles may be determined by folding a sample of the heavy paper and measuring the width of the resulting crease. The pattern for the strip is then obtained by first drawing parallel lines so that the distance between them is equal to the height of your pattern triangle. Next, using the pattern triangle, draw the 5 triangles so that each one is separated from its neighbor by a pair of parallel lines (providing for the crease between them). Score the crease lines firmly so that, when the strip is cut out, it will fold easily and smoothly between the triangles. Fold each pattern piece so that the score lines will be on the inside of the completed model.

The 3 strips for the cube are constructed using the same underlying principles, with one minor additional feature. Begin by determining, experimentally, the length of the *diagonal* of the required square by placing an edge of the completed big tetrahedron along the diagonal of an oversized square. In this way determine the appropriate size for a *pattern square*. From a piece of lightweight paper, cut out one pattern square. Then draw parallel lines on the heavy paper so that the distance between them is equal to the length of the *side* of the pattern square. Then, using your pattern square, draw the 5 squares of Figure 6.1(d), so that each square is separated from its neighbor by parallel scored lines. But—and this is the additional feature to which we referred—in this case you should make the allowance for the crease about *twice* as wide as that for the width between the triangles. This is because the strips of the cube must wrap *around* each other when the model is constructed (see the diagram of the completed cube in Figure 6.1). These pattern pieces should then be cut out and folded along the score lines so that the score lines will be on the inside of the completed model.

Now, just in case you forget some of the real-life details and end up with pattern pieces that don't fit together nicely because there was not enough allowance made for the hinge, there is a way to salvage your effort. The "way out" does not give as good a result as carefully following the original advice, but it is very comforting psychologically. It is to trim off from each edge of the defective piece a small amount, as shown:

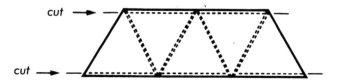

It is not difficult to see that what this does is to truncate at the vertices of the finished model; but if only a small width is trimmed off, the effect on the appearance of the finished model is not noticeable—especially if it is made from brightly colored paper.

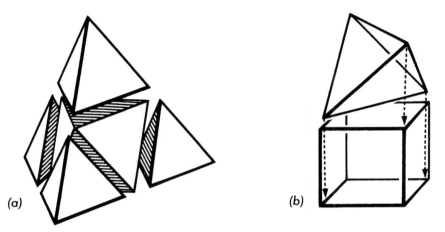

(a)   (b)

Figure **6.2**

Of course, the point of the puzzle is first to figure out how to construct each of the required polyhedra from the specified number of strips and then to get them to fit together as described in the instructions. If you are adventurous, you may wish to try this first on your own. Don't reject this suggestion too quickly; you are very likely to be more successful than you would guess. You can always return to the following instructions later.

6.4   Assembling the Braided Tetrahedron

On a flat surface, such as a table, lay one strip *over* the other strip exactly as shown in Figure 6.3. The fold lines should all be valley folds as viewed from above the table. Think of triangle *ABC* as the *base* of the tetrahedron being formed; for the moment, triangle *ABC* remains on the table. Then fold the bottom strip into a tetrahedron by lifting up the two triangles labeled *X* and overlapping them, so that *C'* meets *C*, *B'* meets *B*, and *D'* meets *D*. Don't worry about what is happening to the top strip, as long as it stays in contact with the bottom strip where the two triangles originally overlapped. Now you will have a tetrahedron, with three triangles sticking out from one edge. Complete the model by carefully picking up the whole configuration, holding the overlapping triangles *X* in position, wrapping the protruding strip around two faces of the tetrahedron, and tucking the *Y* triangle into the open slot along the edge *BC*.

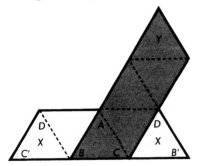

Figure **6.3**

6.5 Assembling the Braided Octahedron

To construct the octahedron, begin with a pair of overlapping strips held together with a paper clip, as indicated in Figure 6.4(a). Fold these two strips into a double pyramid by placing triangle a_1 under triangle A_1, triangle a_2 under triangle A_2, and triangle b under triangle B. The overlapping triangles b and B are secured with another paper clip, so that the configuration looks like Figure 6.4(b). Repeat this process with the second pair of strips, and place the second pair of braided strips over the first pair, as shown in Figure 6.4(c). When doing this, make certain the flaps with the paper clips are oriented exactly as shown. Complete the octahedron by following the steps indicated in Figure 6.4(c). You will note that after step 1 you have formed an octahedron; performing step 2 simply places the flap with the paper clip on it against a face of the octahedron; in step 3, you should tuck the flap *inside the model*.

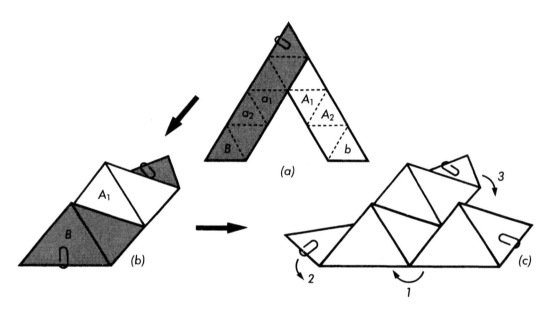

Figure **6.4**

When you become adept at this process you will be able to slip the paper clips off as you perform these last three steps—but this is only an aesthetic consideration, since the clips will be concealed inside the completed model.

After completing the construction of the 4 small tetrahedra and the octahedron, take the models and place a tetrahedron on alternate faces of the octahedron, as shown in Figure 6.2(a). Construct the large tetrahedron containing these polyhedra by placing this configuration on triangle ABC after the two large strips have been put in the position shown in Figure 6.3.

6.6. Assembling the Braided Cube

The cube may be constructed by first taking one strip and clipping together, with a paper clip, the end squares. Then take a second strip and wrap it around the outside of the "cube" so that one square covers the clipped squares from the first strip and the end squares cover one of the square holes. Secure the end squares of the second strip with a paper clip. Make certain that the overlapping squares of the second strip do not cover any squares from the first strip and that the first paper clip is covered. It should now appear as shown in Figure 6.5(a). Now slide the third strip underneath the top square so that two squares of the third strip stick out on both the right and left sides of the cube. Tuck the end squares of this third strip inside the model through the slits along the bottom of the cube, as indicated in Figure 6.5(b). If you now turn the completed cube upside down, it may be opened by pulling up on the strip that covers the top face (this square will be attached inside the model by a paper clip, so you may have to pull firmly) and folding back the flaps that were the last to be tucked inside the model.

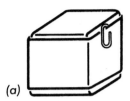
(a)
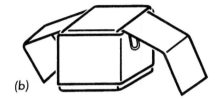
(b)

Figure **6.5**

If you've done this carefully, you can insert the large tetrahedron into the cube and close the cube by tucking the flaps back into their original positions. If you have trouble placing the tetrahedron inside the cube, take another look at Figure 6.2. Once the tetrahedron is placed inside the cube the paper clips are not necessary for holding the faces of the cube in place, because the tetrahedron will exert pressure from inside that will hold the strips in their proper positions.*

A Variation

The given construction of the cube (with a different color on each strip) will yield a cube with opposite faces of the same color, because each strip goes alternately over and under each successive strip it meets as it travels around the cube. There is another construction, using the same strips, that produces a cube in which pairs of adjacent faces are the same color. In this construction each strip goes over two strips and then under two strips as it travels around the cube. Once you have mastered the idea of the first construction, you may wish to assemble your cube strips in this alternative configuration.

REFERENCES

Pedersen, Jean. "Jennifer's Puzzle." *Matimyas Matematika* (1980).
Pedersen, Jean. "The Magic of Reality: An Analogy Between the Star of David and the Stella Octangula." *Matimyas Matematika* (1982).

---

*On most models friction makes the paper clips unnecessary even if the tetrahedron is not inside the cube.

7 Constructing Pop-up Polyhedra

Required Materials

☐ Piece of brightly colored posterboard 22 in. × 28 in.
☐ 6 rubber bands (about 6 in. in circumference)
☐ Yardstick
☐ Ballpoint pen
☐ Scissors
☐ Paper clips
☐ White glue (for paper)

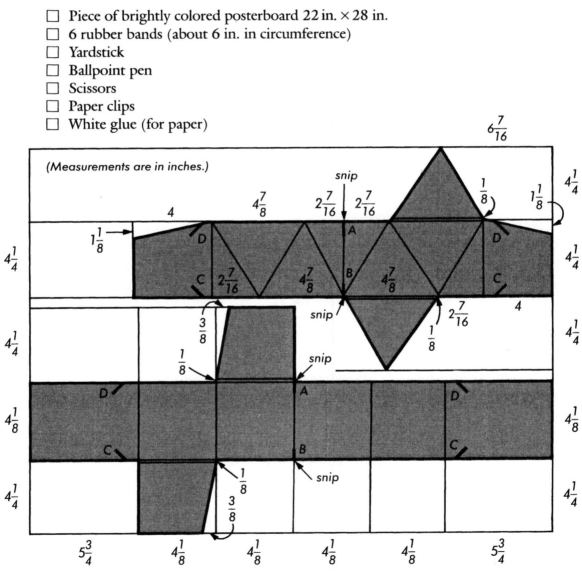

Figure 7.1 Pattern pieces for the octahedron (top) and cube (bottom).

7.1 Preparing the Pattern Pieces

Begin by drawing the pattern pieces on the posterboard as shown in Figure 7.1. Press hard with the ballpoint pen so that the posterboard will fold easily and accurately in the final assembly. Label the points indicated *A*, *B*, *C*, *D*. Be certain to put the labels on what will become the cube or the octahedron when the model is finished—not on the paper that surrounds it. Cut out the pattern pieces and snip the notches at *A* and *B* (but *not* the notches at *C* and *D*).

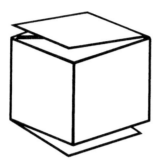

Figure 7.2

7.2 Constructing the Pop-up Cube

1. Crease the cube pattern piece on all the indicated fold lines, remembering that the unmarked side of the paper should be on the outside of the finished cube. Thus each individual fold along a marked line should hide that marked line from view.
2. Position the pattern piece so that it forms a cube with flaps opening from the top and the bottom, as shown in Figure 7.2.
3. Attach the two rectangles together inside the cube with paper clips (this is tempo-rary). Then, with the cube still in its *up* position, cut through both thicknesses of paper at once to produce the notches at *C* and *D*.
4. Connect three rubber bands together, as shown in Figure 7.3.

Figure 7.3

5. Slide one end-loop of this chain of rubber bands through the slot at *A* and the other end-loop through the slot at *B*, leaving the knots on the *outside* of the cube.
6. Stretch the end-loops of the rubber bands so that they hook into slots *C* and *D*, as shown in Figure 7.4(a). The bands must produce the right amount of tension in order for the model to work. If they are too tight the model will not go flat, and if they are too loose the model won't pop up. You may need to do some experiment-ing to obtain the best arrangement.
7. Remove the paper clips when you are satisfied that the rubber bands are perform-ing their function.
8. To flatten the model, *push* the edges labeled *E* and *F* toward each other, as shown in Figure 7.4(b), and wrap the flaps over the flattened portion, as in Figure 7.4(c).

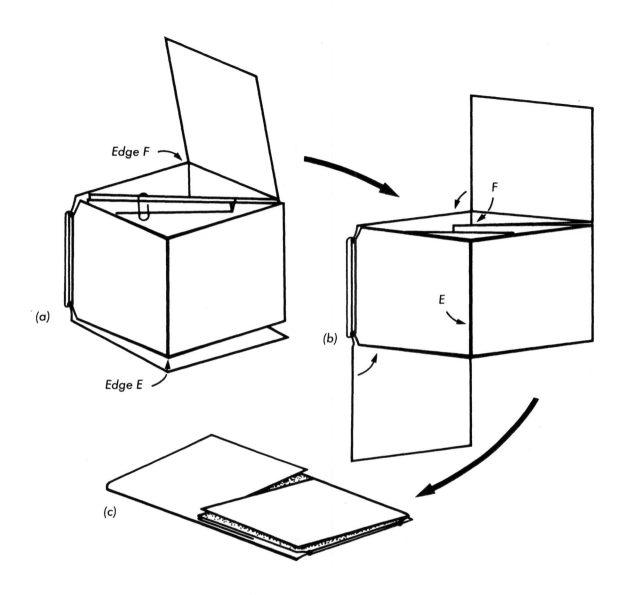

Figure **7.4**

9. Holding the flaps flat, toss the model into the air and watch it POP UP. If you want it to make a louder noise when it snaps into position, glue an additional square onto each visible face of the cube in its *up* position. This also provides the opportunity of adding color to the finished model.

7.3   Constructing the Pop-up Octahedron

1. Crease the octahedron pattern piece on all the indicated fold lines so that the marked lines will be on the inside of the finished model.
2. Position the pattern piece so that it forms an octahedron, with triangular flaps opening on the top and bottom, as shown in Figure 7.5(a). Don't be discouraged by the complicated look of the illustration—the construction is so similar to the cube that once you have the pattern piece in hand, it becomes clear how to proceed.

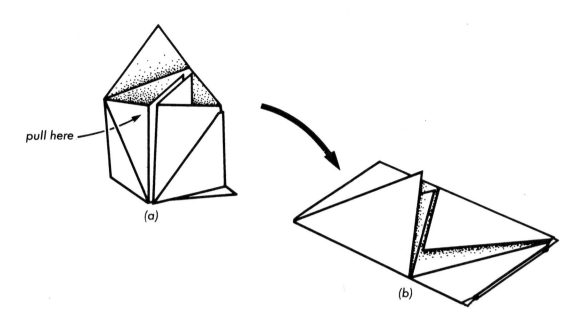

pull here

(a)

(b)

Figure 7.5

3. Secure the quadrilaterals inside the octahedron with paper clips and cut through both thicknesses of paper to make the notches at C and D. Angle these cuts toward the center of the octahedron (so that the rubber bands will hook more securely). Gluing the quadrilaterals inside the model to each other in their proper position produces a sturdier model.
4. Connect three rubber bands together as shown in Figure 7.3.
5. Slide one end-loop of the rubber band arrangement through slot A and the other end-loop through slot B, leaving both knots on the outside of the octahedron.
6. Stretch the end-loops of the rubber bands so that they hook into the slots at C and D. Some adjustment in the size of the rubber bands may be necessary, so experiment to find the best arrangement.
7. Remove the paper clips when you have a satisfactory arrangement of rubber bands.
8. To flatten the model, put your fingers inside and *pull* at the points A and D so that you are pulling those opposite faces *away* from each other until each one is folded along an altitude of that triangular face. Then wrap the triangular flaps over the flattened portion so that it looks like Figure 7.5(b).
9. Holding the triangular flaps flat, toss the model upwards and watch it POP UP. Just as with the cube, this model will make more noise if you glue an extra triangle on each exposed face. This again gives you the opportunity of adding color to the finished model. See Figure 7.6.

NOTE If you store either the cube or the octahedron in its flattened position for several hours or days, it may fail to pop up when tossed in the air. This is because rubber bands temporarily lose their elasticity when stretched continually for long periods of time (a hysteresis effect). If the rubber bands have not begun to deteriorate, the model will behave normally if you let the rubber bands contract for a short while.

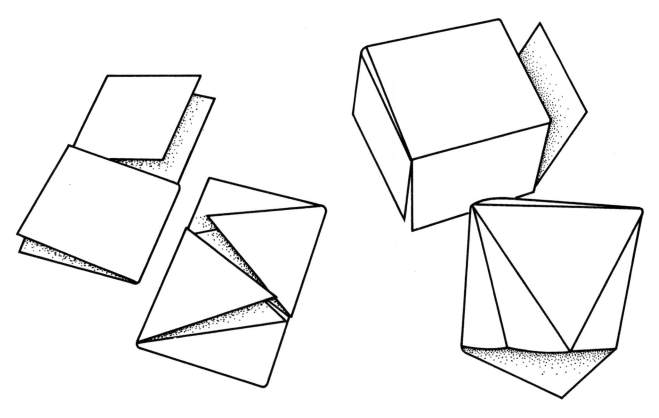

Figure **7.6**   Flattened pop-up polyhedra.   The same polyhedra one second later.

The authors wish to thank Les Lange, the Editor of *California MathematiCs*, for giving permission to use some of the ideas that were originally part of Jean Pedersen's article "Pop-up Polyhedra," *California MathematiCs* (April 1983), pp. 37–41.

8 Constructing Dodecahedra

The faces of the

dodecahedron ————
small stellated dodecahedron ——
great dodecahedron ————
great stellated dodecahedron ——

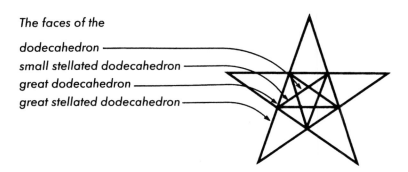

The construction of each dodecahedron described in this section involves the use of gummed mailing tape. In each case the tape must first be folded by the U^2D^2 procedure described in Section 1.4. All the models in this section will have 12 faces, and each face will have 5 sides. It is somewhat surprising that 12 of the regular convex pentagons (Figure 8.1(a)) can interpenetrate each other to form the *great dodecahedron*—but it is even more astonishing that 12 of the *pentagrams* (Figure 8.1(b)) can also interpenetrate each other in two very different ways, to form, in one case, the *small stellated dodecahedron* and, in the other, the *great stellated dodecahedron*. It isn't at all surprising that 12 pentagons with a hole in each (Figure 8.1(c)) can form the framework of an ordinary convex dodecahedron—but, in fact, it is very interesting to construct this model, the *golden dodecahedron*, by simply braiding 6 strips of the U^2D^2 tape (of Section 1.4). The result is very beautiful! However, we should remark that, strictly speaking, these fancy dodecahedra are not polyhedra in the precise sense of the definition of a polyhedron we gave in Section 4.3. Of course, geometers often use a less restrictive definition, by which these models would qualify.

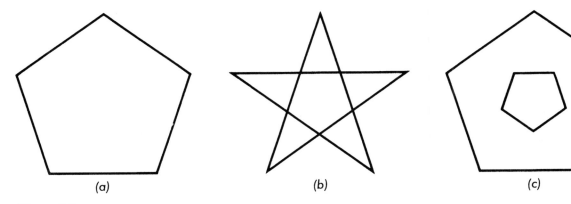

(a)   (b)   (c)

Figure **8.1**

In this chapter we describe, in detail, how to construct each of these models. In each case you will need to know the folding procedure of Section 1.4. We refer to this particular folded tape (the U^2D^2 tape) as a ($\pi/5$)-tape, because the smallest angle on the tape is 36°, or ($\pi/5$) radians (see the final remark of Chapter 2). You may wish to refresh your memory by folding a long strip of this tape and creasing it to produce each of the pentagons shown in Figure 8.2.

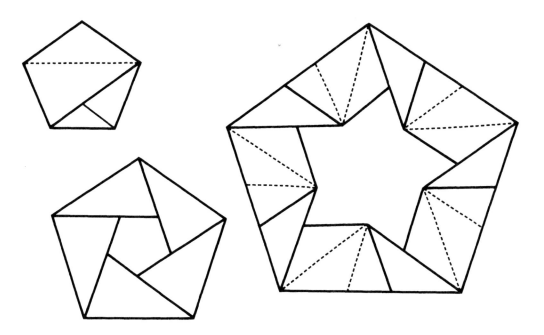

Figure **8.2**   Some pentagons constructed from the ($\pi/5$)-tape

The dodecahedra for which instructions are given next are arranged in ascending order of difficulty of construction. In each case it is assumed that you have an ample supply of gummed mailing ($\pi/5$)-tape.

8.1 The Golden Dodecahedron

Required Materials

- ☐ About 8 yd of gummed mailing ($\pi/5$)-tape (about 2 in. wide)
- ☐ Six different colors of brightly colored wrapping paper. Butcher paper works very well and comes in a variety of colors. Construction paper also works well, but sometimes it must be cut into strips and glued together in order to obtain long enough pieces (unfolded gummed tape may be used to join the strips).
- ☐ Paper clips (at least 30)
- ☐ Scissors
- ☐ Sponge (or washcloth)
- ☐ Shallow bowl
- ☐ Water
- ☐ Hand towel (or rag)
- ☐ Some big (and heavy) books

Preparing the Pattern Pieces

From the folded tape cut 6 strips having 22 triangles each, exactly as shown in Figure 8.3(a). Notice that the cuts occur along the long fold lines.

Select, or prepare, 6 pieces of colored wrapping paper so that each piece is bigger than a single pattern strip. Make certain that the surface on which you are working is large enough to accommodate an entire piece.

Place the sponge (or washcloth) in a bowl. Add water to the bowl so that the top of the sponge is very moist. For each of the 6 strips, moisten one end of the strip by pressing it onto the sponge; then, holding that end, pull the rest of the strip across the sponge. Make certain the entire strip gets wet and then, pulling from both ends, place it gently on the colored paper. Use a hand towel or rag to smooth out the piece (working from the center toward the ends) and wipe up the excess moisture. Make certain every part of the tape bonds to the colored paper.

Put some books on top of the pattern pieces while they dry to keep them flat. When the pattern pieces are dry, cut out each one, trimming off a *very tiny* amount of the gummed tape from the edge (this provides a very neat looking edge). Be careful not to cut off too much, as that may make the finished model unstable. Next refold each piece on the *long* fold lines so that a mountain fold is formed on the colored side of the strip. Leave the short lines *uncreased*.

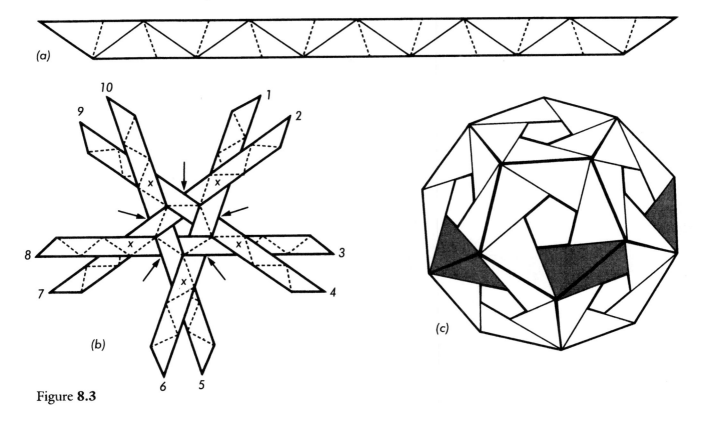

(a)

(b)

(c)

Figure **8.3**

Braiding the Model

To complete the construction of the golden dodecahedron, begin by taking five of the strips and arranging them as shown in Figure 8.3(b), securing them with paper clips at the points marked with arrows. View the center of the configuration as the North

Pole. Lift this arrangement and slide the even-numbered ends clockwise over the odd-numbered ends to form the five edges coming south from the arctic pentagon. Secure the strips with paper clips at the points indicated by crosses. Now weave in the sixth (equatorial) strip, shown shaded in Figure 8.3(c), and continue braiding and clipping, where necessary, until the ends of the first five strips are tucked in securely around the South Pole. During this last phase of the construction, *keep calm* and *take your time!* Just make certain that every strip goes alternately over and under every other strip all the way around the model. When the model is complete, all the paper clips may be removed, and the model will remain stable.

8.2 The Small Stellated Dodecahedron

Required Materials

- ☐ About 6 yd of gummed mailing ($\pi/5$)-tape (about 2 in. wide)
- ☐ Scissors
- ☐ Shallow bowl of water with a sponge (or rag), for moistening the gummed tape when necessary
- ☐ White glue (for paper)
- ☐ Some good books (not merely heavy)
- ☐ Colored paper (optional)

Instructions

We first construct a base dodecahedron and then glue a pentagonal pyramid onto each face. The dodecahedron constructed from tape, as described in the "Alternative Construction" part of Section 4.2 would serve as a suitable base. However, if you do not wish to use those instructions, you may construct the base dodecahedron by another method, which we now describe.

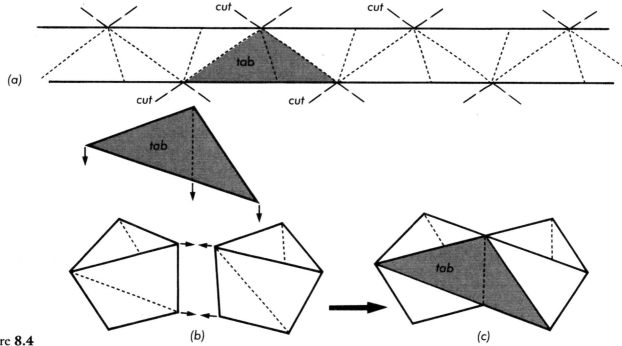

Figure 8.4

First construct 12 pentagons, as shown in step 14 of the instructions of Section 1.4. As each pentagon is constructed, glue all the *overlapping* portions in place. Use a sponge to moisten the appropriate portion of the tape, bending back the parts that are to remain dry so that just the desired parts of the tape come in contact with the sponge. As each pentagon is completed, put it under (or between the pages of) a large book so that it will dry flat.

While the pentagons are drying, cut 30 tabs from the ($\pi/5$)-tape, as shown in Figure 8.4(a). Notice that all the cuts take place along *long* fold lines of the ($\pi/5$)-tape. When the pentagons are dry, begin the assembly by taking two pentagons and a tab, as shown in Figure 8.4(b), and glue the tab across the two pentagons to form one edge of the dodecahedron, as shown in Figure 8.4(c). Complete this phase of the construction by continuing to glue pentagons onto free sides of the existing pentagons on the model (so that there are exactly 3 pentagons around each vertex). When all 12 pentagons have been glued in place, part of the construction will be complete. When you glue on the last pentagon, it is well to proceed by gluing alternate sides into position around the pentagon (rather than consecutive sides)—by doing this any imperfections in the constuction will be more uniformly distributed on the surface of the model (and hence less noticeable).

The final step is to add the stellations to each face of this dodecahedron. Begin this phase by cutting through the small triangles on the ($\pi/5$)-tape as shown in Figure 8.5(a) to form 60 pieces.* Observe that these pieces are not all alike. The (A) pieces

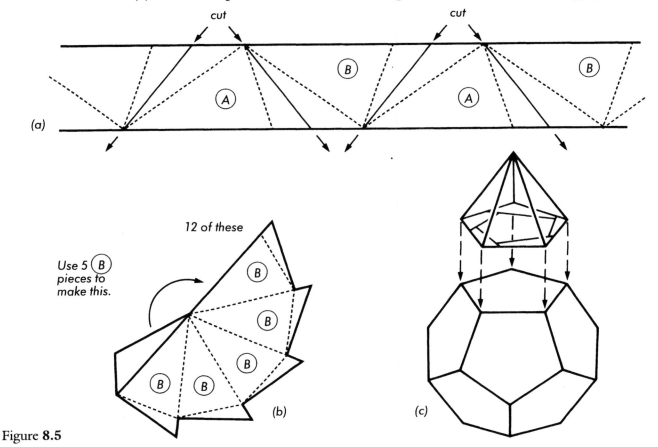

Figure **8.5**

*The angle is not crucial and some deviation is easily tolerated here. Just make a cut that roughly *bisects* the angle at the vertex of the triangle through which you are cutting.

and the (B) pieces, as shown in Figure 8.5(a), are *mirror images* of each other. A pentagonal pyramid may be made by gluing together five (B) pieces, as shown in Figure 8.5(b). When doing this, bend back the tab portion so that you can press just that part against the sponge. Glue the pieces together so that the gummed side of the tape will be inside the finished model. Recrease each of the fold lines along the hinges before joining the last edge (as indicated by the curved arrow of Figure 8.5(b)). Then the tabs around the base of this pentagonal pyramid may be glued to form a platform around the base of the pyramid. The pyramid may then be glued in place, as shown in Figure 8.5(c). This step requires some strong white glue, and for the best results you must hold the pyramid in position until it is well bonded. (You might like to read one of the "good books" while you wait for each pyramid to dry.) Of course, pentagonal pyramids may be made from the (A) pieces as well, and the 60 pieces will provide all the parts for the 12 required pyramids.

The model may be colored by gluing colored pieces of paper to its faces. These pieces may be prepared by first cutting strips of colored paper of the same width as the tape used for the construction and then folding ($\pi/5$)-strips, from which the desired 60 triangles may be cut and glued onto the visible faces of the model. Some craft paper is available with a gummed backing and is particularly easy to use for this purpose. A very attractive coloring is achieved by making all the faces that lie on parallel planes the same colors. This, of course, requires exactly six colors.

8.3 The Great Stellated Dodecahedron

Required Materials

- ☐ About 2 yd of gummed mailing ($\pi/5$)-tape (about 2 in. wide)
- ☐ Some gummed mailing tape that is *not* folded
- ☐ Scissors
- ☐ Shallow bowl of water with a sponge (or rag), for moistening the gummed tape when necessary
- ☐ White glue (for paper)
- ☐ Ruler
- ☐ A good book (not necessarily heavy)
- ☐ Colored paper (optional)

We first construct an icosahedron and then glue a triangular pyramid onto each face. Since the base icosahedron must be constructed from equilateral triangles having an edge length equal to the short line on the ($\pi/5$)-tape, it is necessary to trim off a small amount from the edge of the gummed tape before using it to construct the icosahedron. (Do you see why?) This may be done by beginning to fold the untrimmed tape to produce equilateral triangles (see Section 1.3) and then placing a *short* fold line from the ($\pi/5$)-tape along one fold line of this tape, thereby determining how much needs to be trimmed off (see Figure 8.6). Mark the tape with a ruler and trim off the necessary amount for about 1 yard of tape. Continue folding the equilateral triangles on this trimmed tape. Use this tape to construct the icosahedron according to the instructions given next.

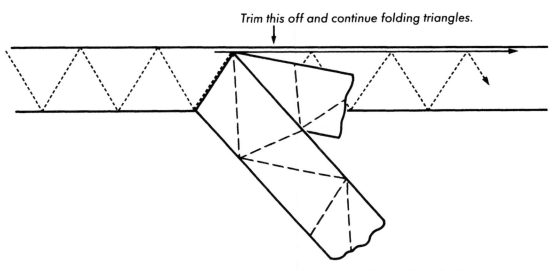

Trim this off and continue folding triangles.

Figure **8.6**   Determining the appropriate width of tape to use for constructing the icosahedron.

Referring to Figure 8.7, first cut a section of 11 equilateral triangles. Glue the first triangle over the last to form the equator of the icosahedron (with the gummed side of the tape on the *inside*). Set that aside and cut a strip of 8 triangles. Take that strip and fold along the lines (1) and (2), shown in Figure 8.7(b), to obtain the hexagon shown in Figure 8.7(c). Glue the overlapping triangles into place and then glue the tab over the triangle next to it, as indicated by the arrow in Figure 8.7(c),

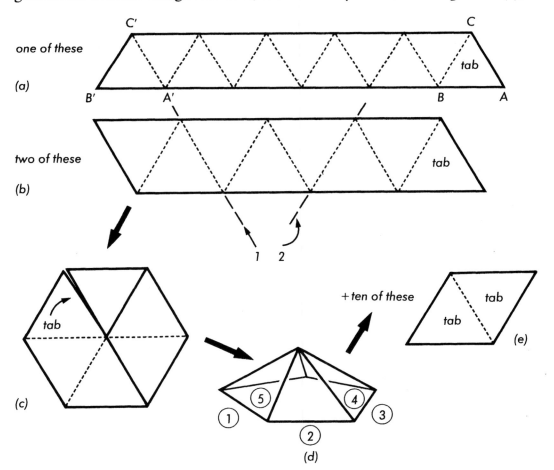

Figure **8.7**

thus producing the baseless pentagonal pyramid shown in Figure 8.7(d). Then cut five 2-triangle tabs, one of which is shown in Figure 8.7(e), and glue one end of each piece onto a face of the pyramid. This pyramid will form the northern (arctic) region of the icosahedron, and it is attached to the equatorial region by gluing into position, in the order designated, the tabs extending from the edges labeled ①, ③, ⑤, ②, and ④. By following this procedure the imperfections (if there are any) will be absorbed around the entire model. The southern (antarctic) region is completed in precisely the same way (in fact, you can simply rotate the model so as to exchange the North and South Poles and repeat the process of covering the arctic region!).

The final step is to add a triangular pyramid to each face of this icosahedron. Begin, as for the small stellated dodecahedron, by cutting through small triangles on the $(\pi/5)$-tape, as shown in Figure 8.8(a). As we have already noted, these pieces will be either Ⓐ pieces or Ⓑ pieces. A triangular pyramid may be made by either gluing together three Ⓑ pieces, as shown in Figure 8.8(b), or by gluing together three Ⓐ pieces to obtain a mirror image of Figure 8.8(b). In either case, after joining the last edge of the pyramid (as indicated by the large arrow in Figure 8.8(b)), snip off the excess from the tabs surrounding the base and glue them into position, as shown in Figure 8.8(c). Each of the 20 pyramids may then be glued into place, one at a time, using a good white glue. You will get a better result if you hold each pyramid in position until it is well bonded. (Relax and read the good book!)

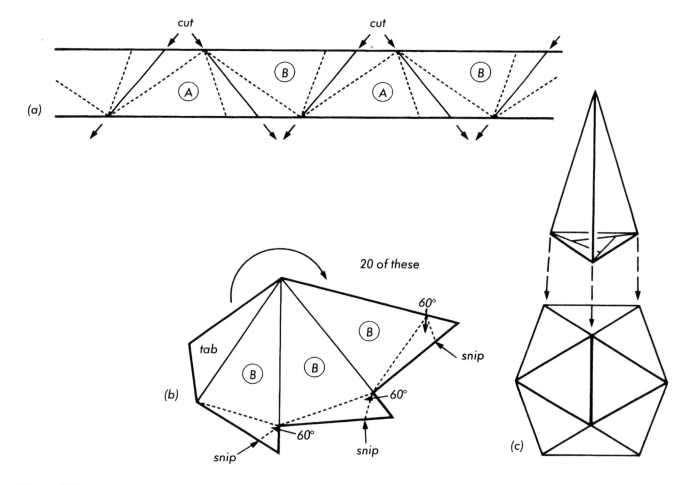

Figure **8.8**

The model may be colored by gluing pieces of paper onto its faces by the same procedure used with the small stellated dodecahedron. Don't throw away the small colored triangles—you will find them perfect for coloring the faces of the great dodecahedron! (See the next construction.)

8.4 The Great Dodecahedron

Required Materials

- ☐ About 2 yd of gummed mailing ($\pi/5$)-tape (about 2 in. wide)
- ☐ Scissors
- ☐ Shallow bowl of water with a sponge (or rag), for moistening the gummed tape when necessary
- ☐ White glue (for paper)
- ☐ Colored paper (optional)

Instructions

This construction is based on the fact that the visible surface of the great dodecahedron may be obtained by replacing each triangular face of the icosahedron by a particular triangular pyramid that points toward the center of the icosahedron.

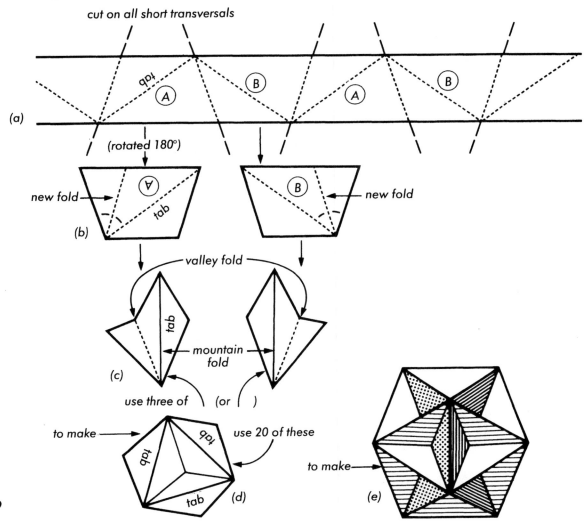

Figure **8.9**

The construction of 20 triangular pyramids from the (π/5)-tape is initiated by first cutting along *short* transversals to produce 60 sections, as indicated in Figure 8.9(a). The sections produced will be of two types, (A) and (B), which are mirror images of each other. Separate the pieces into (A) piles and (B) piles. For each of the 60 pieces make an additional fold line, bisecting one angle of the larger triangle, as indicated in Figure 8.9(b). Make certain each piece is creased so that it has one valley fold and one mountain fold, exactly as indicated in Figure 8.9(c). The triangular pyramids are formed by taking three (A) pieces (or three (B) pieces) and gluing them together to form a baseless triangular pyramid with tabs protruding from each of its three base edges, as shown in Figure 8.9(d). It is important to note that as you look at Figure 8.9(d), the apex of the pyramid is pointing *away* from you and the gummed side of the tape should also be on the side that is not visible.

The construction of the great dodecahedron is now completed by gluing these triangular pyramids together in such a way that (1) the apex of the pyramid points to the center of the polyhedron and (2) the base edges of the pyramids form the edges of a regular icosahedron. You will note as you do this that at every edge there is a choice of placing the tab *on top* or *underneath*. Don't let this worry you; it makes no difference which way you do it. Just proceed *calmly*, remembering that there should be exactly five triangular pyramids around every vertex of the icosahedron. This model is very satisfying to make and you may be surprised at how easily it goes together.

The completed model, shown in Figure 8.9(e), may be colored by gluing colored pieces of paper onto its faces. In fact, if you have already colored either the small stellated dodecahedron or the great stellated dodecahedron and if you saved the pieces you did not use for those models, you will find that they are precisely the pieces you need to color this model.

8.5 *Magical Relationships Between Special Dodecahedra*

It is a curious fact that the visible faces on the golden dodecahedron, the small stellated dodecahedron, and the great stellated dodecahedron are each composed of

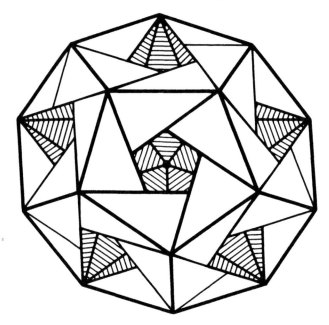

Figure **8.10**

exactly 60 isosceles triangles. Furthermore, if all three of these dodecahedra are constructed from tape of the same width (so that those 60 triangles are all the *same size*), then the small stellated dodecahedron fits inside the golden dodecahedron with the stellations protruding through the holes, touching at the midpoints of the edges of those holes (see Figure 8.10); and the great stellated dodecahedron fits entirely inside the golden dodecahedron with the vertices of both polyhedra coinciding. Is it any wonder that the pentagon and pentagram are associated with magic?

9 Braiding

Platonic Solids

9.1 A Curious Fact

In Chapter 6 we gave instructions for braiding tetrahedra, cubes, and octahedra. The natural question to ask is: **Can we construct the other two Platonic Solids by a similar technique?** In Section 8.1 we described how to braid the skeleton of a dodecahedron, which we called the golden dodecahedron, but we would like to braid a dodecahedron with no holes in its faces. It turns out that we can braid all five of the regular convex solids. In fact, without using any sophisticated mathematics, it is easy to verify the following statement for the five Platonic Solids:

If you make each solid from straight strips of paper fashioned into bands and if you require that all strips on a given model are identical to (or mirror images of) one another, that every edge on the completed model must be covered by at least one strip, that every point on the interior of every face must be covered by at least one thickness from the strips, and that the *same total area* from each strip must show on the finished model,

Then you can braid

the tetrahedron	from	2 strips
the hexahedron (cube)	from	3 strips
the octahedron	from	4 strips
the icosahedron	from	5 strips
the dodecahedron	from	6 strips

(The pattern pieces are shown in Figure 9.1.)

We don't have any general explanation for this curious fact, but we will show you how you can easily demonstrate it. This we do by providing you with instructions for constructing the required polyhedra, and once the polyhedra are constructed, you may then verify, simply by looking at them, that they satisfy the conditions of the preceding statement. To establish the conclusion of the statement, all you need to do is take the models apart and count the number of strips for each one—we call this our "Proof by Destruction." The braided solids are shown in Figure 9.2.

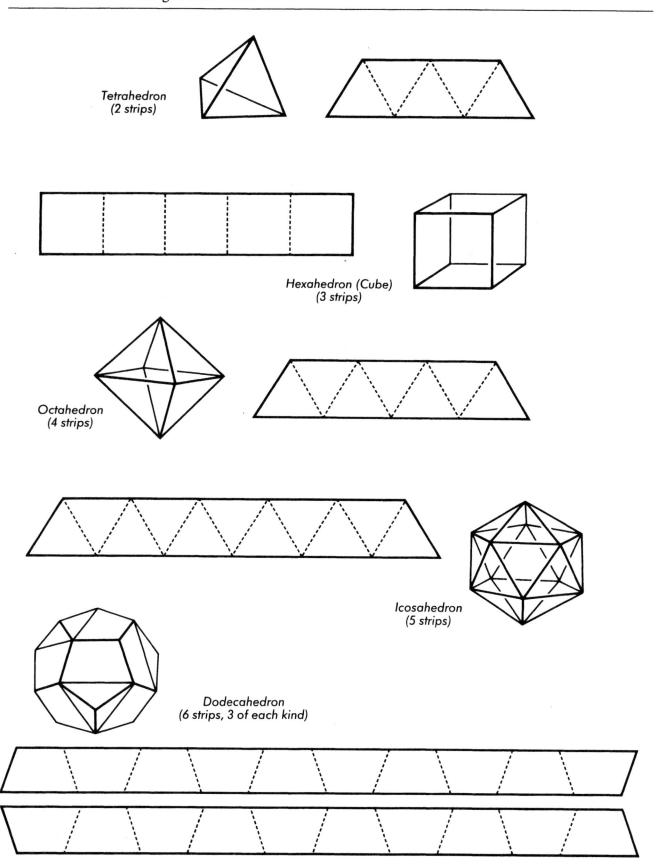

Tetrahedron
(2 strips)

Hexahedron (Cube)
(3 strips)

Octahedron
(4 strips)

Icosahedron
(5 strips)

Dodecahedron
(6 strips, 3 of each kind)

Figure **9.1**

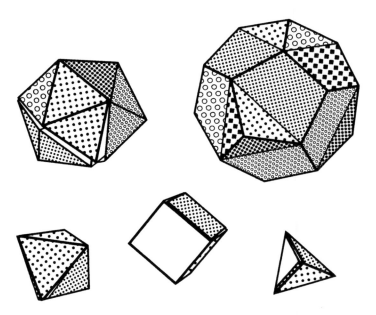

Figure **9.2**   The braided Platonic Solids.

The instructions for how to braid the tetrahedron, cube, and octahedron are in Sections 6.4, 6.6, and 6.5, respectively. What we do in the remainder of this chapter is to explain, first, a general procedure for preparing the strips with folded gummed mailing tape and, second, how to use these strips to braid the icosahedron and dodecahedron in a way that satisfies the hypotheses.

However, we feel we owe it to you to emphasize that this is not the *only* way these polyhedra can be braided from straight strips. Thus, for example, the cube can also be braided with 4 strips instead of 3, still satisfying the conditions of the statement. In order to emphasize this point and to give you an opportunity to see another beautiful model, we give the construction of the *diagonal cube* (braided from 4 strips) in Section 9.3, and then in Sections 9.4 and 9.5 we describe how to braid the dodeca-hedron and icosahedron, respectively.

If you wish to be efficient, we suggest that you look over the six models in this chapter, the golden dodecahedron in Section 8.1, and the rotating ring in Section 10.1, and decide at the outset which ones you want to construct. Then prepare *all* the pattern pieces at once. Having prepared the pieces, try braiding them without our instructions. In case you have any trouble you can always consult the table of contents to find out where the construction is described—but if you don't need any help, you will have the thrill of figuring it all out for yourself. Don't be too modest; you may have more intuition about this than you think. At any rate, you have nothing to lose.

Let's get started.

9.2   Preparing the Strips

Required Materials

☐ Gummed mailing tape; for a sturdier model, use gummed tape that is reinforced with filament. Tape that is about 1½ in. to 2 in. wide is easy to handle for beginners, but if you want larger models, then 3-in. to 4-in. tape is feasible.

- ☐ Paper (preferably colored), onto which you will glue the gummed tape. Butcher paper or gift-wrapping paper is very suitable. Foil wrapping paper can be spectacular, but it sometimes cracks and peels along the creased edges. Experiment with a little piece before making a big investment.
- ☐ Scissors
- ☐ Sponge (or washcloth)
- ☐ Shallow bowl
- ☐ Water
- ☐ Hand towel (or rag)
- ☐ Some heavy books
- ☐ Paper clips
- ☐ Bobby pins (6 will do)
- ☐ Transparent tape (optional)

Having decided which models you want to build, proceed systematically. Observe that the tetrahedron, octahedron, icosahedron, and rotating ring are the only models requiring equilateral triangles. In fact, if you count all the triangles required for these models, you will find it is precisely 181—but it would be tedious and unnecessary to count them as you fold them. It is much easier simply to begin folding the triangles (as in Section 1.3) and then, after throwing away the first few irregular triangles, start cutting off the required pattern pieces; then (when you have just about used up all the folded tape) fold some more triangles, cut off more pieces, . . . until you have all the pattern pieces that require equilateral triangles.

Next construct the pattern pieces for the cube and the diagonal cube. This may be done by using the process described in Section 1.7. Since this is an *exact* process, no part of the tape has to be thrown away. Notice that in order to get the pattern pieces for the cube with 3 strips we only need to use the short lines on the folded tape, but the gummed tape will look like this:

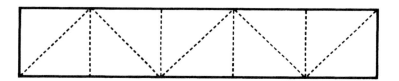

Pattern piece for the cube (requiring 3 strips)

And to get the pattern pieces for the diagonal cube (with 4 strips), we only need to use the long lines on the folded tape, but the gummed tape will look like this:

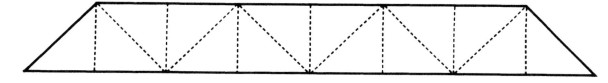

Pattern piece for the diagonal cube (requiring 4 strips)

Finally, prepare the pattern pieces for the two dodecahedra. Notice that both the dodecahedron of Figure 9.1 and the golden dodecahedron require strips that were initially folded by the U^2D^2 process of Section 1.4.

Notice that in the pattern pieces for the dodecahedron of Figure 9.1, we only need to use the short lines on the U^2D^2 (π/5)-tape, but the gummed tape will look like this:

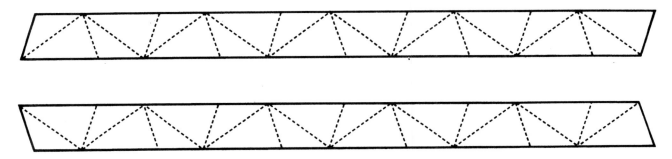

Pattern pieces for the dodecahedron (requiring 3 strips of each kind)

Similarly, notice that to obtain the pattern pieces for the golden dodecahedron of Section 8.1, we only need to use the long lines on the U^2D^2 (π/5)-tape, but the gummed tape will look like this:

Pattern piece for the golden dodecahedron (requiring 6 strips)

Once you have prepared all the gummed tape pieces, get your colored paper and place each of the pieces for a particular model on a different color of paper (to make certain they will all fit). Place a sponge (or washcloth) in a bowl with enough water so that the top of the sponge is very moist or, as we like to say, squishy. Then take each strip (one at a time) and moisten one end by pressing it onto the sponge. Next, holding the moist end, pull the rest of the strip across the sponge. Make certain the entire strip gets wet and then place it on the colored paper. Use a hand towel or rag to wipe up the excess moisture and to smooth the tape into contact with the colored paper. (Only now, when the pattern piece is properly glued to the colored paper, should you worry about cleaning up yourself or the table!)

An efficient scheme is to glue all the pieces that go on one color onto that piece of paper and then go on to the next piece of colored paper. After all the pieces have been glued onto a particular color, put some heavy books on top of the pieces of tape so that they will dry flat. The drying process may take several hours—we can't say, even roughly, how long because it depends on your climate!

When the tape is dry, cut out the pattern pieces, trimming off a *very tiny* amount of the gummed tape from the edge as you do so (this serves to make the model look neater and, more importantly, it allows for the increased thickness produced by gluing the strip to another piece of paper). Then refold each pattern piece so that the raised ridges are on the colored side of the paper. For the models involving equilateral triangles you will refold on every line, but it is *very important* to remember the following rule:

For the	refold only on the
3-strip cube	SHORT lines
4-strip cube	LONG lines
dodecahedron (of Figure 9.1)	SHORT lines
golden dodecahedron	LONG lines

Now you are ready to assemble your models. We strongly urge you first to try it on your own. You may find it useful to clip the beginning and end of each strip together with a paper clip so that you can see how the pieces fit together. As you become more expert you will figure out how to eliminate the paper clips. But if you want more help, turn to the appropriate section for specific instructions.

9.3 Braiding the Diagonal Cube

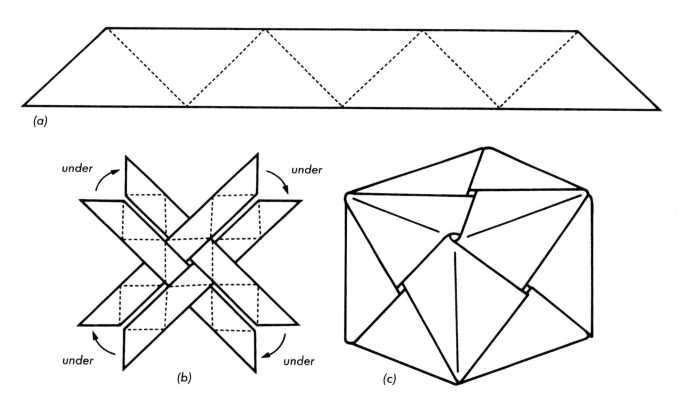

Figure **9.3**

Begin by laying the 4 strips on a table with the colored side *down*, exactly as shown in Figure 9.3(b). The first time you do this it may be helpful to secure the center (where the 4 strips cross each other) with some transparent tape. Now think of the center square in Figure 9.3(b) as the base of the cube you are constructing and note that the strip near the tail of each arrow should go *under* the strip at the head of the arrow (thus the strip near the tail will be on the outside of the model when it crosses the vertical edge of the cube). The procedure for completing the cube is now almost self-evident, especially if you remember that every strip must go alternately over and under the other strips on the model. It may help to secure the centers of the vertical faces with transparent tape as you complete them, but as you become experienced at this construction, you will soon abandon such aids. The final tabs will tuck in to produce the diagonal cube of Figure 9.3(c).

9.4   Braiding the Dodecahedron

Recall that you will need 6 strips as described in Section 9.2. Take two of these strips and cross them as in Figure 9.4.

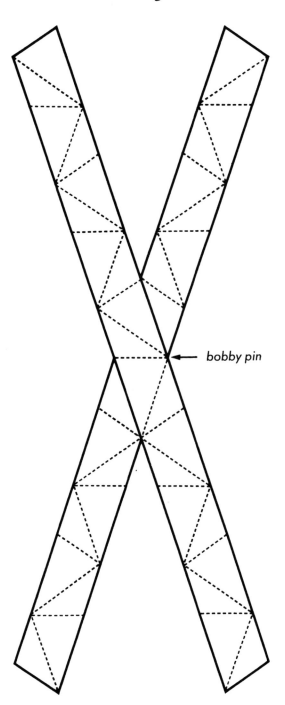

bobby pin

Figure **9.4**

Secure the overlapping edge with a bobby pin, and then make a bracelet out of each of the strips in such a way that

    (1.) four sections of each strip overlap, and

    (2.) the strip that is *under* on one side of the bracelet is *over* on the other side of the bracelet. (This will be true for both strips.)

Use a bobby pin to hold all four thicknesses of tape together on the edge that is opposite the one already secured by a bobby pin.

Repeat the above steps with another pair of strips. You now have two identical bracelet-like arrangements. Slip one inside the other one as illustrated in Figure 9.5, so that it looks like a dodecahedron with triangular holes on four faces.

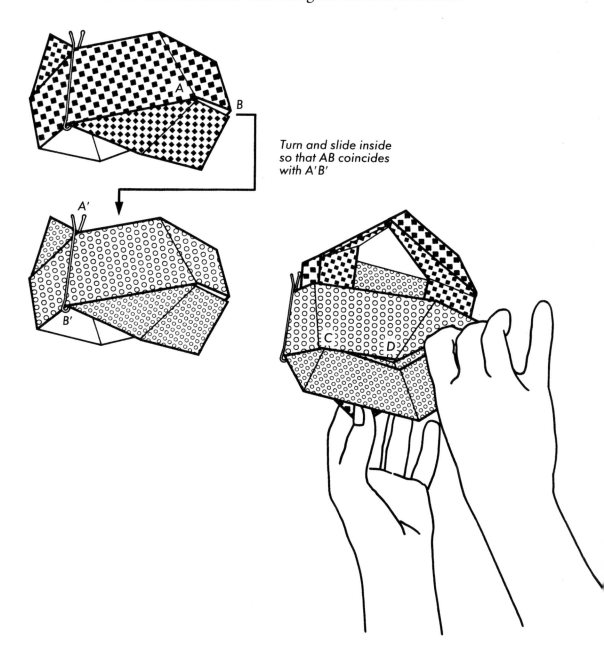

Turn and slide inside
so that AB coincides
with A'B'

Figure **9.5**

Figure **9.5** cont.

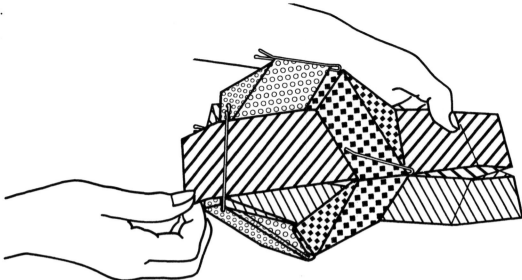

Take the last two strips and cross them precisely as you did earlier (reversing the crossing would destroy some of the symmetry); then secure them with a bobby pin. Carefully put two of the loose ends (either the top two or the bottom two) through the top hole and pull them out the other side so that the bobby pin lands on *CD*. Then put the other two ends through the bottom hole and pull them out the other side. Now you can tuck in the loose flaps, but make certain to reverse the order of the strips—that is, whichever one was on the bottom at *CD* should be on the top when you do the final tucking (and, of course, the top strip at *CD* will be the bottom strip when you do the tucking).

After you have mastered this construction you may wish to try to construct the model with tricolored faces, shown in Figure 9.6. This construction and the one just described are both very similar to the construction for the cube in Section 6.6. The difference is that in the case of the dodecahedron, the three "bracelets" that are braided together are each composed of two strips. This illustrates, rather vividly, exactly how to inscribe the cube symmetrically inside the dodecahedron. To put it another way, it shows how the dodecahedron may be constructed from the cube by placing a "hip roof" on each of the 6 faces of the cube.

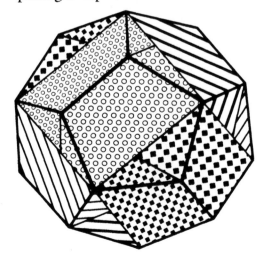

Figure **9.6**

9.5 Braiding the Icosahedron

Label one of the 5 strips with a "1" on each of its 11 triangles; you should write on the side of the paper that will be on the *inside* of the finished model. Then label the next strip with a "2" on each of its triangles, the next with a "3" on its triangles, the next with a "4" on its triangles, and, finally, the last with a "5" on its triangles.

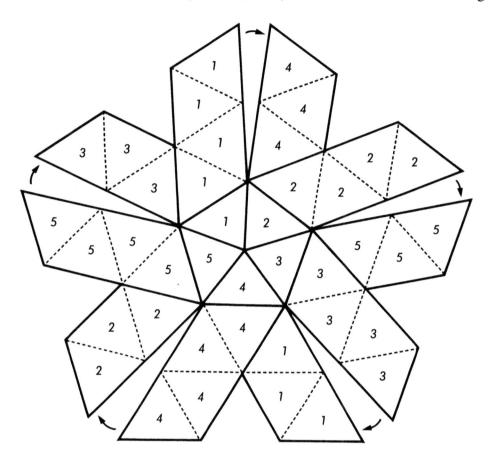

Figure 9.7

Now lay the 5 strips out so that they overlap each other *precisely* as shown in Figure 9.7, making sure that the center 5 triangles form a shallow cup that points *away* from you. You may need to use some transparent tape to hold the strips in this position. If you do need the tape, it works best to put it along the 5 lines coming from the center of the figure (the tape won't show when the model is finished).

Now study the situation carefully before making your next move. You must bring the 10 ends up so that the part of the strip at the tail of the arrow goes under the part of the strip at the head of the arrow (this means "under" as you look down on the diagram; it is really on the outside of the model you are creating, because we are looking at the inside of the model). Half the strips wrap in a clockwise direction and the other end of each of those strips wraps in a counterclockwise direction. What finally happens is that each strip overlaps itself at the top of the model. But, in the intermediate stage, it will look like Figure 9.8(a). At this point it may be useful to put a rubber band around the emerging polyhedron just below the flaps that are sticking out from the pentagon. Then lift the flaps as indicated by the arrows and bring them toward the center so that they tuck in, as shown in Figure 9.8(b).

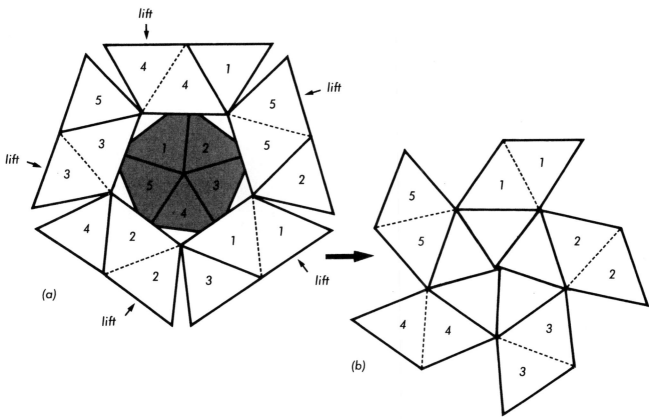

Figure **9.8**

Now simply lift flap 1 and smooth it into position. Do the same with flaps 2, 3, and 4. Complete the model by tucking flap 5 into the obvious slot. The vertex of the icosahedron nearest you will look like Figure 9.9.

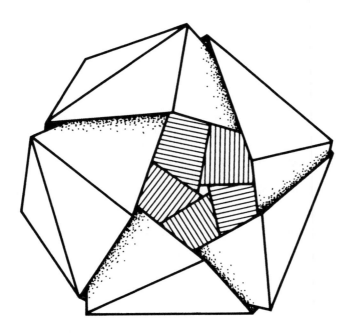

Figure **9.9**

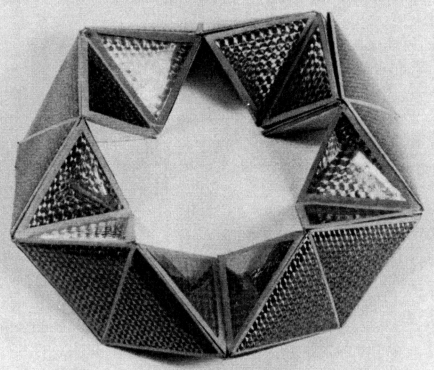

10 Braiding

Rotating Rings

Required Materials

- [ ] About 4 yd of gummed mailing ($\pi/3$)-tape (about 2 in. wide)
- [ ] Two different colors of brightly colored wrapping paper or butcher paper
- [ ] Scissors
- [ ] Sponge (or washcloth)
- [ ] Shallow bowl
- [ ] Water
- [ ] Hand towel (or rag)
- [ ] Some big (and heavy) books
- [ ] Some bobby pins

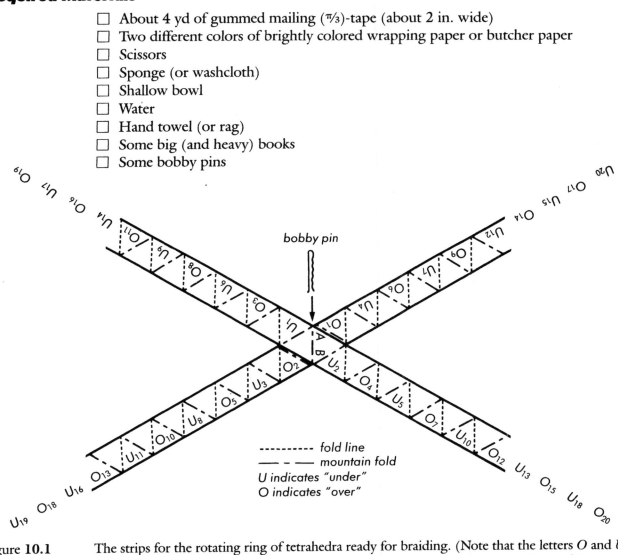

Figure **10.1**   The strips for the rotating ring of tetrahedra ready for braiding. (Note that the letters O and U alternate, and the differences between successive subscripts form a sequence of period 4, namely, ... 1232 ....)

10.1 Braiding a Rotating Ring

Prepare 2 strips, of at least 50 equilateral triangles each, by folding *UD*, as illustrated in Section 1.3. Then glue these folded strips to colored paper and cut them out (see Section 9.2 for practical hints about precisely how to do this to get the best results).

Before beginning the construction of the rotating ring, take each pattern piece and fold the paper, very firmly, in both directions, so that the completed model will flex more easily. You simply cannot overdo this step—in our experience a rotating ring of tetrahedra improves with age due to the increased flexibility of the hinges.

The construction goes as follows:

1. On some fold line near the middle of each strip, cross the 2 strips over each other and secure them together with a bobby pin to form a sideways "X" as shown in Figure 10.1.
2. Label the strips *exactly* as shown.
3. To construct the first tetrahedron, lift the edge *AB* with the bobby pin and slide triangle U_1 underneath triangle O_1; next slide triangle U_2 underneath triangle O_2. Notice that the edges marked — - — will always be edges on the tetrahedron that are not attached to another tetrahedron.
4. Move the bobby pin to the opposite edge of the completed tetrahedron so the strips won't slip apart.
5. Separate the four ends so that U_3, O_3 are on one side and U_4, O_4 on the other, forming a sideways "X" under the tetrahedron like the initial configuration in Figure 10.1.
6. Now repeat the braiding process for the triangles marked U_3, O_3 and U_4, O_4.
7. Again, move the bobby pin to the edge of the new tetrahedron, separate the strips, and repeat the process for triangles U_5, O_5 and U_6, O_6, and so on.
8. When you have braided 10 tetrahedra, trim off all but two triangles from each of the four loose ends at the last edge (which should be secured with the bobby pin).
9. The arrangement of the strips at the end will look like this:

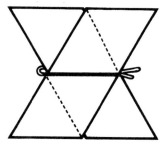

10. Fold back the two triangles, as shown on the left, to get the arrangement on the right.

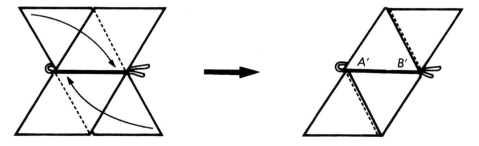

11. Place this edge next to the edge *AB* on the first tetrahedron and tuck the two triangles into the openings (or slots) on the edge of this first tetrahedron so that they go in the backwards direction. Figure 10.2 shows how the last triangle should look as it slides into the slot.

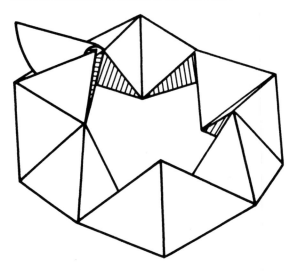

Figure **10.2**

12. Figure 10.3 shows the finished rotating ring, which consists of 10 tetrahedra.

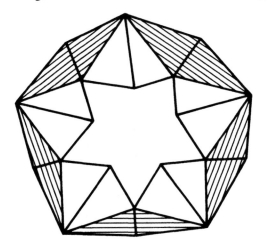

Figure **10.3**

Practical Hints

A common mistake is to fail to cross the strips properly while braiding the model. The result is that, although the model looks fine, it comes apart as you rotate it. If that happens, undo it and rebraid it. A good rule to remember is that when two strips meet at a crossing,

the strip that came from underneath goes over next, and
the strip that was on top goes underneath next.

When you are flexing a new rotating ring of tetrahedra, the paper may have a tendency to buckle. Be gentle with it and push the offending edge back into its proper position. After a while two things will happen. First, you will become more adept at turning the model (because you will have gained a better understanding of its mechanical properties) and, second, the model itself will become more pliable.

10.2 Variations

We were not obliged to use exactly 10 tetrahedra. In fact, a rotating ring can be constructed of just 8 regular tetrahedra (but not 6—try it and you'll see why). We chose to use 10 tetrahedra because we like the symmetry of the final model.

Rings containing 22 or more tetrahedra can be tied in various knots before joining the first and last tetrahedra to each other. These rotating knots form interesting configurations whose twisting motions are almost hypnotic.

Why do you suppose it was possible to make this model? Observe that the regular tetrahedron is the only one of the Platonic Solids whose opposite edges, when extended, do *not* form parallel lines. In fact, it is this property that makes it possible to use these tetrahedra to construct a ring that will rotate. You can produce nonrigid rings by joining the opposite edges of other kinds of polyhedra (cubes, for example), but they will not rotate unless the opposite edges used for joining the polyhedra lie on nonparallel lines. Moreover, if the pair of joining edges on each of several similar polyhedra are not on lines that are at right angles to each other in space, the symmetries of the final rotating ring may be quite unusual.

We suggest that you experiment with different numbers of tetrahedra in your rotating rings. Try tying knots before joining the first and last tetrahedra and, if you are really interested, make some rotating rings with other polyhedra. Figure 10.4, which resembles a holiday wreath in appearance, is made from 14 hexacaidecadeltahedra. (You shouldn't need the hyphens now!)

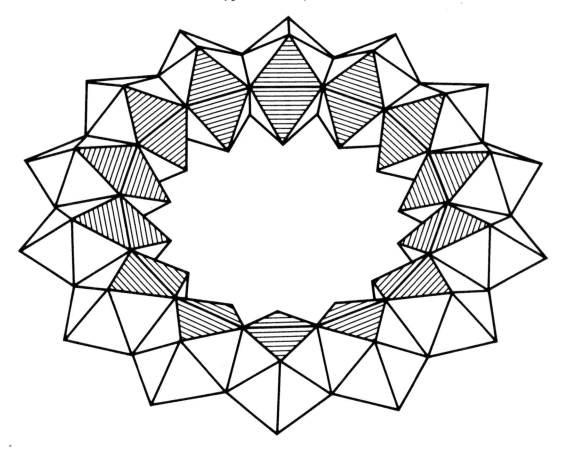

Figure **10.4**

Some More Fun with Your Rings

Suppose you wish to number the visible faces of your rotating ring of 10 tetrahedra with the consecutive numbers 1, 2, 3, ..., 40. Now $1 + 2 + 3 + ... + 40 = 820$. A very nice way of seeing this is to rearrange the sum as follows and add vertically:

$$\begin{aligned}
1 + \ \ 2 + \ \ 3 + ... + 19 + 20 + \\
40 + 39 + 38 + ... + 22 + 21 \\
\hline
41 + 41 + 41 + ... + 41 + 41
\end{aligned}$$

We readily see that 41 occurs precisely 20 times. Thus,

$$1 + 2 + 3 + ... + 40 = 20(41) = 820$$

Now, since 820 is exactly divisible by 10 (the number of tetrahedra in our ring), it is sensible to ask whether or not it is possible to number the faces of the rotating ring with the numbers from 1 to 40 so that the numbers on the four faces of each tetrahedron will sum to 82.

We believe you will be able to answer this question—and that you will also be able to ask and answer the corresponding question for rings involving other numbers of tetrahedra. It is reported in *Mathematical Recreations and Essays* by Rouse Ball and Coxeter (referenced at the end of this chapter) that a certain R. V. Heath in answering the corresponding question for a rotating ring of 8 tetrahedra showed how one could assign the numbers 1, 2, 3, ..., 32 (which sum to 528) to the 32 faces so that the four faces of each tetrahedron sum to 66, and "corresponding" faces, one from each tetrahedron, sum to 132.

Suppose you consider a rotating ring of N tetrahedra. The sum of the integers 1, 2, 3, ..., $4N$ is $2N(4N + 1)$. Our original question has to do with assigning one of these integers to each face so that the sum for each tetrahedron is $2(4N + 1)$. Heath's refinement would require that the sum of "corresponding" faces be $\frac{N(4N + 1)}{2}$; plainly to achieve this refinement N must be even.

Doris Schattschneider and Wallace Walker have produced a monograph (see references) that includes die-cut nets from which fascinating solids and rotatings rings can be constructed that have Escher-type designs on their faces. Their kit would surely give you many ideas for decorating not only rotating rings but many other models that you have made from this book.

REFERENCES

Pedersen, Jean J. "Braided Rotating Rings." *Mathematical Gazette*, 62 (1978), pp. 15–18.

Rouse Ball, W. W., and H.S.M. Coxeter. *Mathematical Recreations and Essays*, 12th ed. Toronto: University of Toronto Press, 1974.

Schattschneider, Doris, and Wallace Walker. *M. C. Escher Kaleidocycles*. Corte Madera, CA: Pomegranate Artbooks, 1987.

11 Constructing

Collapsoids

11.1 What Is a Collapsoid?

There is an interesting class of polyhdra having the property that all faces are congruent parallelograms. Since all faces are parallelograms, the polyhedra in this class have the property that every edge determines a *zone* of faces such that each face in the zone has two sides equal and parallel to the given edge. Polyhedra having this latter property are called *zonohedra*; we may speak of an *n*-zonohedron to emphasize that the polyhedron in question has *n* zones. As interesting examples of polyhedra in this class, the *rhombic dodecahedron* (which has 12 faces and 4 zones) and the *rhombic triacontahedron* (which has 30 faces and 6 zones) appear in Figure 11.1, which is based on illustrations by H.S.M Coxeter.

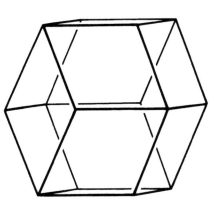 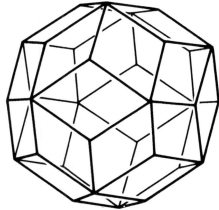

Rhombic dodecahedron (n = 4)   Rhombic triacontahedron (n = 6)

Figure **11.1**

In *Regular Polytopes*, Coxeter describes the general theory of zonohedra and states that the angles on the faces of the rhombic dodecahedron are 70°32′ and 109°28′, while the angles on the faces of the rhombic triacontahedron are 63°26′ and 116°34′. You will readily believe that these angles are *not* ones that we get easily by folding paper (though we could get them in principle!). However, these are beautiful models and we can construct polyhedra like them by replacing each of the rhombic faces with a four-faced pyramid without its base, which is composed of four equilateral triangles.

137

We call this a *cell* and refer to each of the triangles as a *triangular face* of the cell. One of the authors (Jean Pedersen) experimented with such cells in the hope that the flexibility of the cell might make it possible to approximate the rhombic faces (see the references at the end of this chapter).

The experiment showed that the desired models cannot be made with each pyramid projecting out from the polyhedron's center. However, when each pyramid projects in toward the center of the polyhedron, you get a *pseudo-zonohedron*. Furthermore, the models turn out to have a very surprising property apparently not possessed by the real zonohedra, namely, they *fold up and lie flat* (but see a remark in Section 11.5). All we have to do is leave unattached a sequence of edges that go from any vertex to the vertex diametrically opposite it. This very surprising and pleasing feature was first discovered by the two children of one of the authors (then six and nine years old) while they were playing with the partially constructed models (as their mother was preparing supper).

Because these pseudo-zonohedra can fold up flat in various ways, we have named them *collapsoids—polar* if they collapse about an axis between two poles, and *equatorial* if they collapse about a zone (which may be thought of as an equator). In the pages that follow we give you step-by-step instructions for constructing and collapsing these models. Then, in Section 11.5, we suggest some investigations you might want to make for yourself.

11.2 Preparing the Cells, Tabs, and Flaps

Required Materials

- ☐ Gummed mailing tape; for sturdier models, use gummed tape that is reinforced with filament. Any width between 1½ in. and 3 in. will be easy to handle.
- ☐ Scissors
- ☐ Sponge (or washcloth)
- ☐ Shallow bowl
- ☐ Water
- ☐ Hand towel (or rag)
- ☐ Colored paper, preferably with a self-adhesive backing (optional)

Begin by taking the gummed tape and folding a strip of 50 or more equilateral triangles (as shown in Section 1.3). Leave the folded tape attached to the roll so that you can fold more triangles as you need them. Observe that the new triangles you fold will become more and more accurate as long as you don't cut off the last triangle and start again from scratch. Remember to cut off and discard the first few irregular triangles. Once you have the process started you can cut off from the tape the number of triangles (specified in Section 11.3) required to construct the cells, tabs, and flaps as described in Section 11.3.

Each of the four collapsoids discussed in this chapter requires a certain number of *cells*, *tabs*, and *flaps*, which are described next. A table at the beginning of Section 11.3 tells you precisely how many cells, tabs, and flaps are required for each of the four collapsoids whose construction is outlined in that and the following section. We suggest that you look through Sections 11.3 and 11.4 and decide which model, or models, you would like to make, then construct all of the required parts, and finally turn to the directions that tell you how to glue those parts together.

Cells

Each cell is constructed from a straight strip of 6 equilateral triangles that look like this:

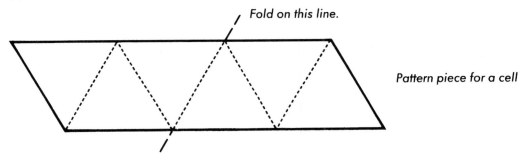

Fold on this line.

Pattern piece for a cell

Fold this strip as indicated and glue the overlapping portions together (if the sticky sides are not together, fold the paper in the other direction).* It should look like Figure 11.2.

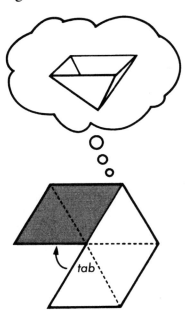

tab

Figure **11.2**   Becoming a cell.

As the bubble in Figure 11.2 indicates, this piece really aspires to be a "baseless" pyramid. To achieve such a pyramid, overlap the triangle labeled *tab* with the triangle indicated by the arrow. Once you see that your result looks like that shown in the bubble, glue the overlapping triangles together.

We call this baseless pyramid a *cell*. Notice that each cell may be pressed flat in two directions. As you make each cell lay it flat, first in one direction and then in the other, and while it is flat, crease the two edges very firmly. Then place the cells on a table to dry. You may wish to stack them on top of each other in piles that make it easy for you to keep track of how many you have constructed (say, 5 or 6 to a pile).

*Moistening the required triangles may be done by patting the gummed side of the triangle *that will be glued* against a moist sponge.

Tabs

Tabs, those pairs of triangles that will be used to connect the cells together, are the easiest parts to construct—you simply cut off sections containing two triangles each.

A word of caution is required here, however. We should remind you that, because of the way the tab fits on the completed model, the hinge should be creased so that the sticky sides of the tape come together (see Figure 11.3).

When you have made the tabs, stack them in a pile.

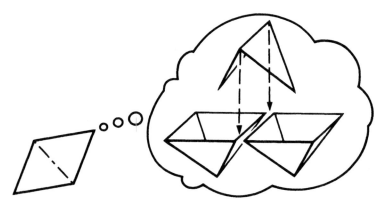

Figure 11.3   A tab doing its work.

Flaps

The purpose of the flap, which is a tab that only gets attached to a cell by one of its triangles, is to hold the model together when it is expanded. Flaps allow us to open up some edges of the polyhedron and fold it flat for storing. Since the flap will stay in place only if it is fairly stiff, we need to make it sturdier. Here is one way to do it.

Begin with a 3-triangle piece of tape and a 2-triangle piece of tape. Glue one triangle of the 2-triangle piece to the center triangle of the 3-triangle piece as shown in Figure 11.4(a). Make certain the sticky sides of the tape are facing you. Then wrap the end triangles of the 3-triangle piece over the center triangle, as shown in Figure 11.4(b), and glue them in place, as shown in Figure 11.4(c). Press the three thicknesses flat and crease the remaining edge firmly. Stack the flaps in a pile separate from the tabs.

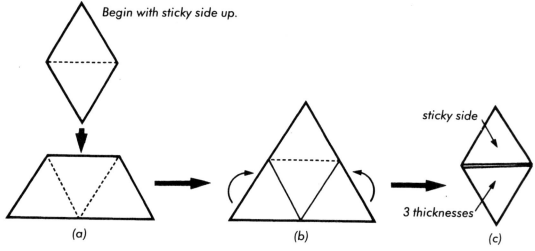

Figure 11.4   Constructing a flap.

11.3 *Constructing 12-, 20-, and 30-Celled Polar Collapsoids*

Collapsoid	Number of cells	Number of tabs	Number of flaps
12-celled polar	12	20	4
20-celled polar	20	35	5
30-celled polar	30	54	6
12-celled equatorial	12	18	6

Constructing the 12-Celled Polar Collapsoid

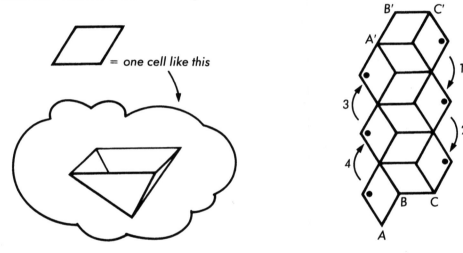

Figure **11.5**

Figure 11.5 represents a net of baseless pyramids. Each rhombus in the figure signifies one cell. Begin the construction by joining 6 cells together in a string. They will look like this:

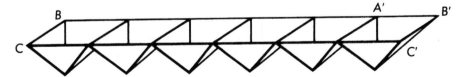

Then join the cells on the right and left so that, as you look down on the figure, (1) you are looking into each cell and (2) the outline of the cells looks precisely like the net in Figure 11.5.

At this point it may help you to label the triangles at the head and tail of each arrow with an identifying number (as suggested by the numbers next to the curved arrows). Then use the tabs to join together the cells with like numbers. The vertices of the cells labeled with heavy dots will then be next to each other. You will notice that an edge going from one heavy dot to $A(A')$ to $B(B')$ to $C(C')$ and ending at the other heavy dot will remain open.

Your model should now look like Figure 11.6.

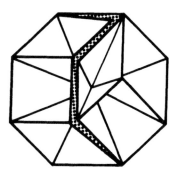

Figure **11.6**   The 12-celled collapsoid, without flaps.

At this point we suggest you try collapsing your model. It should fold flat in the shape of ⁴⁄₆ of a regular hexagon. Press it fairly gently into the flattened position and bring it back into its expanded shape several times so that you get the feel of the mechanical motion. Now all that remains is to attach the flaps.

Flaps should be attached to provide a covering for the four open edges. One very effective way is to attach flaps alternately to one or other of the loose sides along the open edge. More precisely, think of the edges as labeled 1, 2, 3, and 4 as you traverse from North Pole to South Pole on this model; then attach flaps to the left-hand side on edges 1 and 3 and to the right-hand side on edges 2 and 4. The effect of this is that the flaps *interlock* and hold the model together better than they would if all the flaps had been attached to the same sides of the open edge.

Practical Hint

It may happen that, as you complete the model by sliding the flaps into place, you observe that a triangular face of the cell into which you want to tuck the flap seems a little flimsy. If so, glue another tab around this face. The result will be a very sturdy cell into which you can now tuck the flap. This hint is useful for making any of the collapsoids (see Figure 11.7).

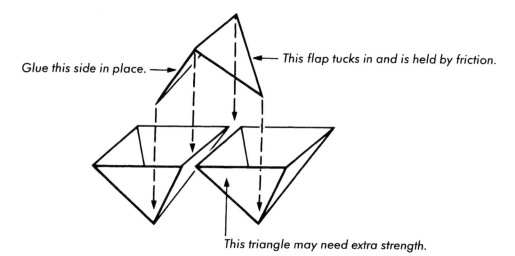

Glue this side in place. ⟶   ⟵ *This flap tucks in and is held by friction.*

This triangle may need extra strength.

Figure **11.7**   A flap in position.

Constructing a 20-Celled Polar Collapsoid

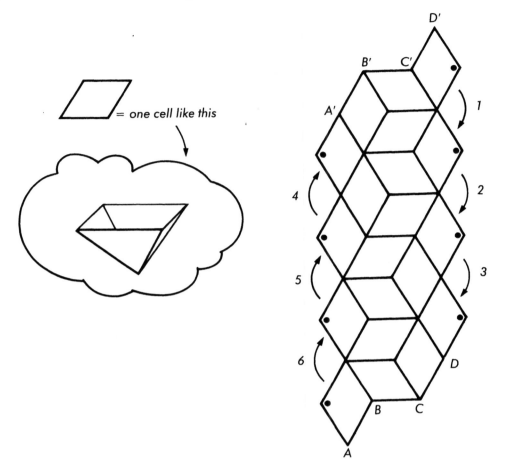

= one cell like this

Figure **11.8**

As before, Figure 11.8 represents a net of baseless pyramids. Begin this construction by joining 8 cells together in a string. These are the cells in the zone going from $B'C'$ to BC in the net diagram. Then continue by joining the cells on the right and left so that, as you look down on the figure, (1) you are looking into each cell, and (2) the outline of the cells looks precisely like Figure 11.8.

If you feel it would be helpful, label the triangles at the head and tail of each arrow with an identifying number (as shown in Figure 11.8). Then use the tabs to join together the cells with like numbers. The vertices of the cells labeled with heavy dots will then be next to each other. You will notice that an edge going from one heavy dot to $A(A')$ to $B(B')$ to $C(C')$ to $D(D')$ and ending at the other heavy dot will remain open.

Collapse the model into ⅚ of a regular hexagon and bring it back into expanded position several times until you understand the mechanics of its motion.

Attach the flaps alternately to one or the other of the loose sides along the open edge. Think of the edges as labeled 1, 2, 3, 4, and 5 as you traverse from North Pole to South Pole; then attach flaps to the left-hand side on edges 1, 3, and 5 and attach flaps to the right-hand side on edges 2 and 4.

You may need to reinforce the triangular faces onto which the flaps fit in the cells, as described in the earlier Practical Hint.

Constructing a 30-Celled Polar Collapsoid

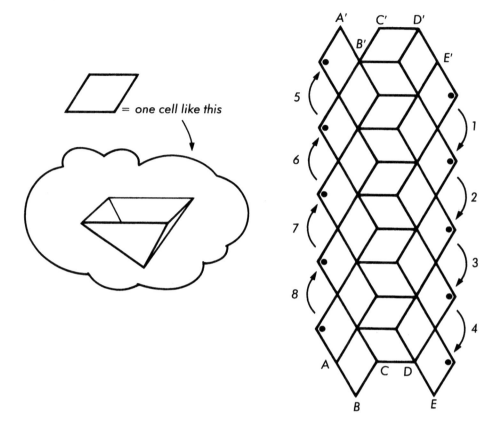

= one cell like this

Figure 11.9

Figure 11.9 represents a net of baseless pyramids. Begin this construction by joining 10 cells together in a string. These are the cells that go around the model from $C'D'$ to CD. Then join the cells on the right and left of this zone so that as you look down on the figure you are looking into the cells, and the outline of the cells looks precisely like Figure 11.9.

Next, the sides of the cells at the head and tail of the arrows should be joined to each other (so that the vertices of the cells bearing a heavy dot come together). The edge going from one heavy dot to $A(A')$ to $B(B')$ to $C(C')$ to $D(D')$ to $E(E')$ and ending at the other heavy dot will remain open. This model collapses into the shape of a complete regular hexagon.

Flaps may be attached to the open edge in the same alternating fashion as described for the 12- and 20-celled collapsoids. That is, think of the open edges as though they were labeled 1, 2, 3, 4, 5, and 6 as you traverse from North Pole to South Pole on this model; then attach flaps to the left-hand sides on edges 1, 3, and 5 and to the right-hand sides on edges 2, 4, and 6. As before, it may be necessary to reinforce a triangular face of a cell before tucking in the flap.

11.4 Constructing the 12-Celled Equatorial Collapsoid

Figure 11.10(a) represents a net of baseless pyramids. Each parallelogram in the net represents one cell. Begin the construction by joining 6 cells together in a string. These are the cells that go along one zone (the equator of this model) between $B'C'$ and BC.

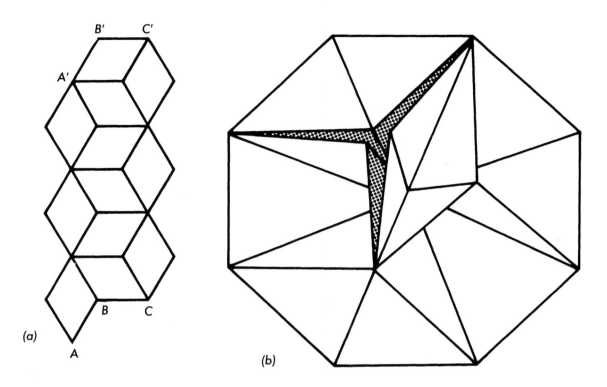

Figure **11.10**

Next join the cells on the right and left so that as you look down on the figure you are looking into the cells, and the outline of the cells looks precisely like the net in Figure 11.10(a).

Now join the side $B'C'$ to BC and then the side $A'B'$ to AB. You now have a 12-celled collapsoid, as shown in Figure 11.10(b), that will fold flat about the ring of 6 cells forming the equatorial zone.

Flaps added to the three cells on either side of the equatorial zone will help to keep the model in its inflated form. Make the flaps all go in either a clockwise or a counterclockwise direction about each pole. And, as with the other models, you may wish to reinforce the triangle onto which the flap falls when it is in its final position.

11.5 Challenges

Now that you have constructed your polyhedra, you should get to know them. If you color them in various ways you will learn a great deal about their symmetries and how they are related to other polyhedra you have constructed.

One way to color the models is to get some gummed colored paper, available at art stores or office supply stores, and prepare a number of 2-triangle tabs in assorted colors. These may then be glued on top of the faces you wish to color. If gummed paper is not available, ordinary colored paper may be glued onto the faces.

When we talk of coloring an edge, we mean gluing a colored tab over that edge. The effect of this gluing will be that two adjacent triangles on the surface of the collapsoids are colored.

We now give you some specific suggestions for coloring.

For any collapsoid

Color the zones: Color one convex edge* red, for example; then color the edge opposite that edge red, and the edge opposite that edge red, . . . until you have colored all of the edges in that zone. Then begin again on any uncolored edge and repeat the process with another color. Repeat two or three more times and you will have colored all edges (and, hence, all faces!). You will then be able to see the zones of the collapsoid very clearly.

For the 12-celled collapsoid (polar or equatorial)

Color the cube: About a vertex where 4 convex edges come together, color each of those edges red, for example, and also color red the 4 convex edges surrounding the diametrically opposite vertex. Then begin again at any other uncolored vertex surrounded by 4 convex edges and color those edges blue, for example, and also color blue the 4 convex edges surrounding the opposite vertex. There will remain just two diametrically opposite pairs of vertices surrounded by 4 uncolored convex edges. Color those 8 edges with a third color. Compare this model with a cube!

Color the octahedron: About a vertex where 3 convex edges come together, color each of those edges red, for example, and also color red the 3 convex edges surrounding the diametrically opposite vertex. Then begin again and repeat the process with another color. Do this two more times. Compare this model with an octahedron!

For the 30-celled collapsoid

Color the dodecahedron: About a vertex where 5 convex edges come together, color each of those edges red, for example; then color red the 5 edges surrounding the diametrically opposite vertex. Repeat this process, using a new color each time, until you have colored all 60 convex edges. Compare this model with a dodecahedron!

Color the icosahedron: About a vertex where 3 convex edges come together, color each of those edges red, for example; then color red the 3 edges surrounding the diametrically opposite vertex. Repeat this process, using a new color each time (you will need 10 colors), until you have colored all 60 convex edges. Compare this model with an icosahedron!

Other Equatorial Collapsoids (for the expert)

Both the 20- and the 30-celled collapsoids can be made in an equatorial form—but not with equilateral triangles. To see this, notice that the 20-celled collapsoid has 8 cells in each zone, and the 30-celled collapsoid has 10-cells in each zone. Thus the 20-celled equatorial collapsoid must be made from cells that are parts of an *octagon*, and the 30-celled equatorial collapsoid must be made from cells that are parts of a *decagon*. In each case, when you construct the model with the appropriate cells, you proceed as

---

*By a convex edge, we mean an edge that would be in contact, along its entire length, with a tight elastic material sheathing the collapsoid. Alternatively, we call an edge of a collapsoid *convex* if we can rest the collapsoid on a flat surface with that edge touching the surface.

before. The only difference is that, just as in the case of the 12-celled equatorial
collapsoid, you ignore the arrows and, instead, connect the ends of the principal zone
(the one you first constructed). In this way you get a flowerlike arrangement about
both poles, which may be held shut with paper clips; the entire model will collapse
about the equatorial zone. Since you know (from Chapter 1) how to fold both 8-
gons and 10-gons, we may confidently leave the exploration of these models to our
very enthusiastic readers! See Figures 11.11 and 11.12.

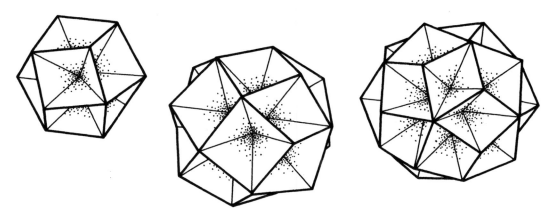

(a) 12-, 20-, and 30-celled collapsoids.

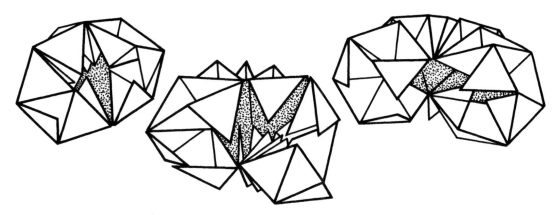

(b) Partly collapsed.

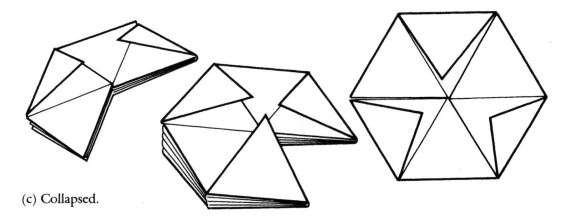

(c) Collapsed.

Figure **11.11**

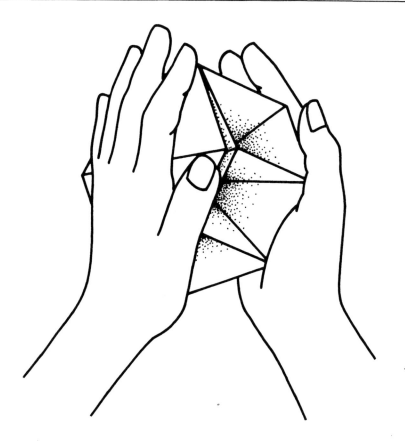

(a) 12-celled equatorial collapsoid, partly collapsed.

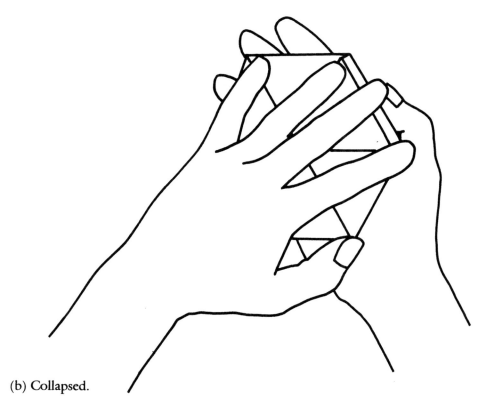

(b) Collapsed.

Figure **11.12**

Other Polar Collapsoids

There are many other polar collapsoids. Perhaps the most spectacular one is the 90-celled one, which may be colored to bring out its relationship with the dodecahedron (or the icosahedron). A brief description of how to construct it appears in Jean Pedersen's article "Collapsoids." (See the references at the end of this chapter.)

Another Very Easily Constructed Collapsoid

You may have realized that the ordinary regular hexahedron, or cube, is a zonohedron. So it is natural to ask: **Will the cube collapse if we make it like our other collapsoids?** The answer is both yes and no. It is impossible to replace each face with one of our cells (why?), so in this respect we get a negative answer. However, it is possible to construct the cube from a special net (as in Section 4.2) on which each face has been folded along the diagonal lines before we assemble it. If such a cube is constructed and then cut apart along the line indicated in Figure 11.13, it *will* collapse in the expected manner. Try it!

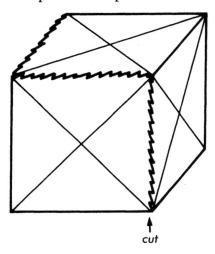

cut

Figure **11.13**

Of course, this brings up another question: **Would the rest of the ordinary zonohedra collapse if they were made from faces that were scored along both diagonals and then cut apart along a line following an edge from one vertex to its diametrically opposite vertex?** (The answer is *yes!*)

REFERENCES

Coxeter, H.S.M. *Regular Polytopes*, 2nd ed. New York: Macmillan Mathematics Paperbacks, 1963.
Pedersen, Jean J. "Collapsoids," *Mathematical Gazette*, 59 (1975), pp. 81–94.

12 Some Mathematics of Polyhedra

In this final chapter we give you some of the mathematics associated with the poly-hedra you constructed in the previous chapters. In fact, we give you a very small part of the mathematics, but enough, we hope, to convince you of the mathematical richness of the ideas with which you have become familiar. We hope to leave you with the awareness that the models you have constructed are not only aesthetically pleasing but also an excellent source of mathematical ideas. We want to emphasize that it is no coincidence that the study of polyhedra can play this second role, too. For mathematics is the only science capable of fulfilling the function of elucidating and analyzing the remarkable phenomena around us and explaining why those phenomena exhibit the features that we notice. Thus there is always some mathematics to be discovered in anything that commands our attention; and by subjecting these phenomena—cosmic rays, weather patterns, the spread of contagious diseases, Platonic Solids—to mathematical analysis, we enrich our appreciation of their significance, and thereby improve our control of our environment.

Thus, while we would understand and sympathize if you decided to skip this chapter, we would advise against it. We're sure you would regret it if you never made the effort to understand the mathematics contained in the fascinating models you have so conscientiously constructed.

12.1 Mathematical Applications of Jennifer's Puzzle

Volumes of Some Related Polyhedra

Recall that Jennifer's puzzle was the subject of Chapter 6. We will first show how the puzzle may be used to calculate some volumes of Platonic Solids. We start from the following three facts about volumes:

> The volume of a rectangular parallelepiped of sides a, b, c is abc.
> The volume of a pyramid of height h standing on a base B is
> $$\frac{1}{3}h \times (\text{area of } B)$$
> If the linear dimensions of a figure are multiplied by d, the volume is multiplied by d^3.

Now, it follows immediately from the first fact that the volume of a cube of side a is a^3. It is perfectly possible to use the second fact to compute the volume of a regular tetrahedron of side a to be $\frac{a^3}{6\sqrt{2}}$. However, this calculation requires us to calculate the area of an equilateral triangle of side a—which is $\frac{\sqrt{3}a^2}{4}$—and the height of the regular tetrahedron—which is $\sqrt{\frac{2}{3}}a$. There is, in fact, an easier method provided by thinking of a regular tetrahedron placed inside the cube, as in Jennifer's puzzle. The empty space inside the cube and outside the regular tetrahedron then consists of four congruent tetrahedra; and if the regular tetrahedron has side a, the box has side $\frac{a}{\sqrt{2}}$, as can be seen from Figure 12.1. Moreover, each of the four tetrahedra is an upright wedge, as shown in Figure 12.2.

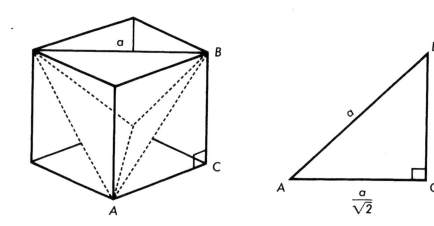

Figure **12.1**

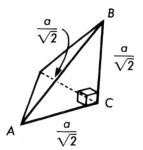

Figure **12.2**

Thus the volume of each wedge is, according to the second fact,

$$\frac{1}{3}\frac{a}{\sqrt{2}}\left(\frac{1}{2}\left(\frac{a}{\sqrt{2}}\right)^2\right) = \frac{a^3}{12\sqrt{2}}$$

Since the volume of our original regular tetrahedron (which we call $V_T(a)$) is equal to the volume of the cube minus the volume of the four wedges, we see that it may be calculated as

$$V_T(a) = \left(\frac{a}{\sqrt{2}}\right)^3 - 4\left(\frac{a^3}{12\sqrt{2}}\right) = \frac{a^3}{2\sqrt{2}} - \frac{a^3}{3\sqrt{2}} = \frac{a^3}{6\sqrt{2}}$$

Notice that this is the same as the result we claimed earlier. Notice, too, that we have proved the following:

The volume of a regular tetrahedron is $\frac{1}{3}$ of the volume of the smallest cubical box it sits in.

For, as we calculated, our box had a volume of $\frac{a^3}{2\sqrt{2}}$.

It is now very simple to compute the volume of the regular octahedron of side a. For, as observed in Jennifer's puzzle, we may place regular tetrahedra on four of the faces of the regular octahedron to obtain a regular tetrahedron of side $2a$. Thus if $V_T(a)$ is the volume of the regular tetrahedron (of side a) and $V_O(a)$ is the volume of the regular octahedron (of side a), then

$$V_O(a) + 4V_T(a) = V_T(2a) \tag{V}$$

Then, using the known value for V_T, we obtain

$$V_O(a) + 4\frac{a^3}{6\sqrt{2}} = \frac{(2a)^3}{6\sqrt{2}}$$

yielding

$$V_O(a) = \frac{4a^3}{6\sqrt{2}} = \frac{\sqrt{2}a^3}{3}$$

Notice that this last argument also shows the following:

The volume of the regular octahedron of side a is 4 times the volume of the regular tetrahedron of side a.

Despite this result you would be doomed to failure if you tried to *construct* a regular octahedron of side a by putting together four regular tetrahedra of side a—it just can't be done! The volume measures are the same, but that is not enough. We observe, however, that to prove our last result it is quite unnecessary to calculate the volume of the tetrahedron; we simply apply (V), knowing that by doubling the length of each edge on any polyhedron, the volume of that polyhedron is increased by a factor of 2^3, or 8. And this, of course, we know by virtue of the third fact in our original list of facts about volumes.

However, if we put together the two results displayed (in boxes) earlier, we obtain a third interesting comparison:

The volume of the regular octahedron of side a is $\frac{1}{2}$ of the volume of the regular tetrahedron of side $2a$ and hence $\frac{1}{6}$ of the volume of the cube in which the octahedron sits with its vertices at the midpoints of the faces of the cube.

If we glue a regular tetrahedron onto *each* face of a regular octahedron (of side *a*), we obtain a figure called the *Stella Octangula* by the great astronomer Johannes Kepler (see Figure 12.3). This may be thought of as two interpenetrating regular tetrahedra of side 2*a*, intersecting in the regular octahedron. These facts allow us to deduce immediately that the volume of the Stella Octangula is 12 times the volume of the regular tetrahedron of side *a*, or 3 times the volume of the regular octahedron of side *a*. Even more interesting, we have the next result:

> The volume of the Stella Octangula is $\frac{1}{2}$ that of the smallest cube into which it can be placed.

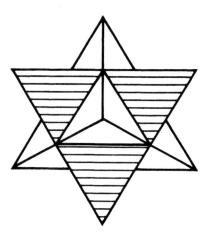

Figure **12.3**   Stella Octangula

Symmetries of a Cube

We next consider the *symmetries* of a cube. After talking (rather a lot) about this important geometrical concept, we will again go back to Jennifer's puzzle to see how it casts light on the relation of the symmetries of a cube to the symmetries of a regular octahedron and those of a regular tetrahedron.

We picture the cube occupying a certain part of space; by a *symmetry* we mean the effect of a rotation of the cube about its center that brings it into a position occupying the same original part of space. Thus, for example, we may rotate the cube through 90° about an axis passing through the midpoints of two opposite faces; this is a symmetry of the cube. It is plain that:

1. if we follow one symmetry by another, the composite effect is again a symmetry,
2. if we reverse a symmetry we again get a symmetry, and
3. the "zero" rotation, that is, the "rotation" that holds every point fixed, is trivially a symmetry.

These three facts allow us to talk of the *group* of symmetries of the cube (or, more generally, of the *group* of symmetries of any polyhedron). Notice that a symmetry is completely determined when we describe the position of the points of the cube after the rotation—it is thus sufficient to describe the destinations of each vertex.

Now, a classical way to study the group of symmetries of the cube is to look at the **4 *main diagonals*** of the cube, that is, the 4 straight-line segments that pass from a vertex of the cube to the diametrically opposite vertex. It is plain that any symmetry of the cube permutes these 4 main diagonals, in the sense that, on executing the rotation, some main diagonal (perhaps the original one) comes to occupy the position in space originally occupied by any given main diagonal (see Figure 12.4 for an example).

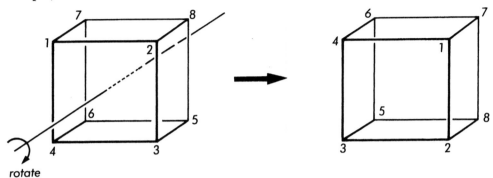

Rotation through 90° about the axis joining the midpoints of front and back faces.

$$\text{Diagonal} \left.\begin{array}{c} 15 \\ 26 \\ 37 \\ 48 \end{array}\right\} \text{ moves into the position originally occupied by diagonal} \left.\begin{array}{c} 26 \\ 37 \\ 48 \\ 15 \end{array}\right\}$$

Figure **12.4**

The following notation is often used to specify a permutation. We refer to the main diagonals 15, 26, 37, and 48 as D_1, D_2, D_3, and D_4, respectively. Then the permutation described in Figure 12.4 may be written

$$\begin{pmatrix} D_1\, D_2\, D_3\, D_4 \\ D_2\, D_3\, D_4\, D_1 \end{pmatrix}$$

indicating that

D_1 moves into the position originally occupied by D_2.

D_2 moves into the position originally occupied by D_3

⋮

However, an even more convenient notation for this permutation (which we adopt in Section 12.4) is

$$(D_1 D_2 D_3 D_4)$$

This is the *cyclic* notation; you should think of the symbols D_1, D_2, D_3, D_4 written, in that order, around a circle, and the notation indicates that each element (diagonal) is replaced, in the permutation, by its successor in the given cyclic order. This permutation is, then, a *cycle of length* 4.

Let us give another example. The permutation written in the more cumbersome notation as

$$\begin{pmatrix} D_1\, D_2\, D_3\, D_4 \\ D_3\, D_4\, D_1\, D_2 \end{pmatrix}$$

appears, in cyclic notation, as

$$(D_1 D_3)\, (D_2 D_4)$$

for D_1 "goes to" D_3, which goes to D_1, and D_2 goes to D_4, which goes to D_2. This permutation is thus a composition of 2 cycles, each of length 2. Every permutation can be written in cyclic notation as a composition of cycles. If some element is fixed under the permutation (for example, a rotation about a main diagonal fixes that diagonal), we think of that element as constituting a cycle of length 1. The permutation that moves nothing is called the *identity*; in cyclic notation it is $(D_1)(D_2)(D_3)(D_4)$.

We have seen then that every rotation of the cube induces a permutation of the set of 4 main diagonals. It is less obvious that given any permutation of the 4 main diagonals, there is exactly one symmetry of the cube that effects this permutation. Let us give you just one key argument leading to this important conclusion. Let us ask: **What symmetry could transform each main diagonal into itself?** If such a symmetry sends vertex 1 to vertex 1, we claim it must send vertex 2 to vertex 2. For, if not, it sends vertex 2 to vertex 6, and this is impossible because 12 is an edge of the cube and 16 is not. Likewise it must send vertex 3 to 3 and vertex 4 to 4. In other words, it is the zero movement (or rotation). Thus it follows that the only nontrivial symmetry leaving the diagonals alone, if it existed, would have to send vertex 1 to vertex 5, vertex 2 to vertex 6, vertex 3 to vertex 7, and vertex 4 to vertex 8.

We are going to use the idea of *orientation* to show that there is no such symmetry. If we orient the faces of the cube so that *opposite* orientations are induced in the common edge of two faces—this is called "orienting the cube"—and if we orient the face 1234 by $\overrightarrow{1234}$, then we must orient the face 2358 by $\overleftarrow{2358}$, so that we induce opposite orientations in their common edge 23. Likewise, we must orient the face 5678 by $\overleftarrow{5678}$. Figure 12.5 shows this orientation of the cube (one of the two possible orientations). Precisely, it shows the faces of the cube of Figure 12.4 drawn in net form with the orientation of each face indicated by the circular arrows. Notice that an orientation of the cube is determined by the orientation of any one face. It is now plain that 1234 cannot be moved by a symmetry to 5678, since a rotation must preserve orientation; so that there is *no* nontrivial symmetry leaving the 4 main diagonals alone. A consequence of this is that distinct symmetries must produce distinct permutations of the main diagonals.

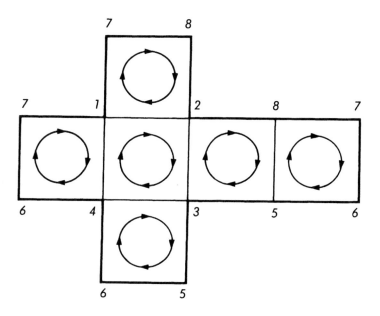

Figure **12.5**

It is not difficult to see that there are precisely $4! = 4 \cdot 3 \cdot 2 \cdot 1 = 24$ permutations of 4 objects. And there are also precisely 24 symmetries of the cube! You should experiment with one of the cubes you have constructed. We suggest that you first use it to verify that a cube really does have 24 symmetries. One way to do this is to suppose you have a cube with each face colored a different color—for the sake of discussion, say the colors are red opposite blue, green opposite orange, and white opposite purple. Further, suppose your cube lives in an imaginary cubical drawer that is only slightly larger than the cube. Figure 12.6 shows how it might look when you open the drawer.

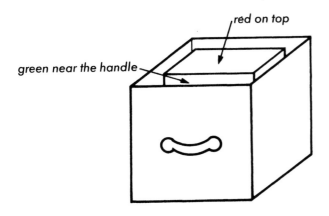

Figure **12.6**

Now, if you want a particular face, say red, to show at the top of the drawer, then there are exactly 4 ways you can achieve this, since you can then have green, orange, white, or purple against the handle of the drawer. (Why can't blue be against the handle of the drawer?) Once you have decided which color you want at the top of the drawer and which color is against the handle, the position of the cube will be completely determined. Do you see why? Since we may choose the top face in 6 ways and then choose the front face in 4 ways, it follows that the cube can be set down in the drawer in exactly 24 different ways.

It now follows that the symmetries that produce the 24 possible positions of the cube in the drawer really do correspond to the 24 permutations of the 4 main diagonals of the cube. You will find it very instructive to complete the table in Section 12.4, which describes each symmetry geometrically (for example, a rotation about the axis through the centers of the faces 1234 and 5678 so that 1 moves to 2, 2 moves to 3, ...). The very important zero rotation, which is also called the *identity*, is listed last in that table. Naturally, it induces the *identity* permutation of the vertices and the main diagonals. The force of our earlier arguments is that the group of symmetries of the cube is *isomorphic* to (that is, structurally equivalent to) the group of permutations of 4 objects, usually written S_4.

The table and accompanying exercise in Section 12.4 contain even more information about symmetry. For example, if, in addition to the data about how the vertices and the main diagonals are permuted, you record the number and length of the various types of cycles in these permutations, you may be able to observe a very interesting pattern that is part of a much bigger picture.

Symmetries of the Regular Octahedron and Regular Tetrahedron

There is a very nice geometrical argument that now allows us to determine the symmetries of the regular octahedron; for, as you will recall from Jennifer's puzzle, a regular octahedron can fit inside a cube with its vertices at the midpoints of the faces of the cube. Thus every symmetry of the cube will send the octahedron into the space it originally occupied within the cube.

But it is also true that the midpoints of the faces of a regular octahedron are the vertices of a cube. Thus every symmetry of the octahedron will send this cube into the space it originally occupied within the octahedron.

It therefore follows that the symmetry group of the regular octahedron is the same as (or isomorphic to) that of the cube. You may be interested to know that the full group of all permutations of n objects is called the *symmetric group of n objects* and is written S_n. Thus, as we have shown, the group of symmetries of the cube and hence also of the octahedron is S_4, which is, of course, a group having 24 elements. In fact, S_4 is also known as the *octahedral* group.

It is particularly revealing to relate the symmetries of the cube to the *diagonal cube* you constructed in Section 9.3; for you constructed the diagonal cube with 4 strips of paper—and these strips are objects that are permuted by each rotation of the cube! Thus the abstract notion of the permutation of the 4 main diagonals becomes much more vivid when you think of the 4 strips that form the surface of the diagonal cube as the objects that are being permuted.

Now let us again return to Jennifer's puzzle. In particular, let us suppose you place the big tetrahedron inside the cube with the vertices of the tetrahedron at the vertices 2, 4, 5, 7 of the cube. Then consider the symmetries of the cube. We claim the following: A symmetry of the cube *either* permutes the vertices 2, 4, 5, 7 among themselves (in which case it also permutes the vertices 1, 3, 6, 8 among themselves) *or* it exchanges each of the vertices 2, 4, 5, 7 for one of the vertices 1, 3, 6, 8. Moreover, exactly half of the symmetries of the cube fall into the first class and exactly half into the second. You can verify these facts in a number of ways. If all else fails, use the table of symmetries of the cube that appears in Section 12.4.

It now follows that the group of symmetries of the regular tetrahedron is precisely the same as the subgroup of the group of symmetries of the cube consisting of those symmetries that lie in the *first* class described earlier. As we pointed out, this group has half of the elements of S_4, that is, 12 elements. In fact, this is the subgroup of S_4 called the *alternating group*, written A_4 (see the reference at the end of this chapter for further details). It is also known, particularly among geometers, as the *tetrahedral* group.

An equivalent way to identify the subgroup A_4 is to ask the question: **Which permutations of the main diagonals of the cube move vertex 2 to one of the vertices 2, 4, 5, 7 and which move it to one of the vertices 1, 3, 6, 8?** Once again, those in the first class constitute the symmetries of the regular tetrahedron.

Remark on Orientation and Symmetry

In our study of the symmetries of the cube, you will remember that we used a rather sophisticated argument about the orientation of the cube (to show that no symmetry of the cube could send each vertex of the cube to the diametrically opposite vertex).

We would like to clarify this argument for you by describing the analogous situation for the symmetries of an equilateral triangle (regular 3-gon):

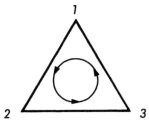

If we allow only *planar* symmetries of the triangle (that is, rotations that take place in the plane of the triangle), then all we can do is to rotate through $\pm 120°$ about the center of the triangle—or, of course, to carry out the zero rotation. Thus the only permutations (in our first cumbersome notation) of the vertices are the following three:

$$\begin{pmatrix} 1\ 2\ 3 \\ 1\ 2\ 3 \end{pmatrix} \qquad \begin{pmatrix} 1\ 2\ 3 \\ 2\ 3\ 1 \end{pmatrix} \qquad \begin{pmatrix} 1\ 2\ 3 \\ 3\ 1\ 2 \end{pmatrix}$$

It is obvious that these all maintain the orientation of the triangle. However, if we allow a rotation in space (of 3 dimensions), then, by rotation through 180° about the line joining vertex 1 to the midpoint of side 23, we may achieve the permutation

$$\begin{pmatrix} 1\ 2\ 3 \\ 1\ 3\ 2 \end{pmatrix}$$

which, as you may observe, *reverses* the orientation of the triangle. Thus in talking about the symmetries of the regular 3-gon (and a similar remark applies to the symmetries of the regular *n*-gon), we should specify whether we insist on planar symmetries or allow rotations in 3-dimensional space.

Now comes the crucial point! In talking of the symmetries of the cube, we should also specify whether we confine our rotations to the 3-dimensional space of the cube (as we have, in fact, done) or allow rotations in a 4-dimensional space! In the former case, orientation is preserved, but it is not necessarily preserved in the latter. It is a tribute to our awareness of the 3-dimensional world in which we live that we adopt (mathematically) different conventions in discussing symmetries of planar and non-planar figures. In fact, by rotation in 4-dimensional space, we can achieve the apparently impossible symmetry of the cube described in this section. Do you see how?

12.2   Euler's Formula and Descartes' Angular Deficiency

Let us look at the five Platonic Solids and record, for each of them, the number of vertices (V), edges (E), and faces (F). The result is the following table:

	V	E	F
Tetrahedron	4	6	4
Hexahedron (cube)	8	12	6
Octahedron	6	12	8
Dodecahedron	20	30	12
Icosahedron	12	30	20

We notice that, in all cases, we have the formula

$$V - E + F = 2$$

This formula is called *Euler's Formula* for polyhedra, and Euler produced arguments to show that the formula holds for *any* polyhedron in our sense (as defined in Chapter 4). For example, for the pentagonal dipyramid, we have

$$V = 7, \qquad E = 15, \qquad F = 10, \quad \text{and} \quad 7 - 15 + 10 = 2$$

as promised by Euler's Formula.

We will not prove the formula here—a logically satisfactory proof is rather difficult—but we will show you, by means of a modified version of an argument due to George Pólya, that Euler's Formula is equivalent to a very deep result, due to the French mathematician and philosopher of the sixteenth century, René Descartes.

Consider any of the convex polyhedra you have constructed in the course of studying this book. If you consider all the faces that come together at a particular vertex and lay them out flat, they will leave a gap. Thus, for example, for the regular tetrahedron, we would get, at any vertex, this picture:

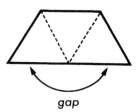

gap

This leaves a gap of 180° at this vertex. For the cube we would get, at any vertex, this picture:

gap

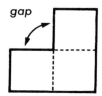

which leaves a gap of 90° at this vertex. In fact, it was Euclid who pointed out (see Section 4.3) that there would always be such a (positive) gap for any convex polyhedron. Descartes called this gap the *angular deficiency* of the polyhedron at the particular vertex. Let us number the vertices of a given polyhedron and write δ_n for the angular deficiency of the polyhedron at the nth vertex. However, we prefer to measure the angular deficiency in radians rather than degrees (remember, π radians = 180°), so that, for example, the angular deficiency at each vertex of a regular tetrahedron is π, and the angular deficiency at each vertex of a cube is $\frac{\pi}{2}$.

Descartes studied the *total angular deficiency* of a convex polyhedron, that is, the sum of the angular deficiencies at each vertex. Let us write Δ for the total angular deficiency so that

$$\Delta = \delta_1 + \delta_2 + \cdots + \delta_v = \sum_{n=1}^{V} \delta_n$$

He proved the remarkable fact that, for *any* convex polyhedron,

$$\Delta = 4\pi$$

Again, we will not attempt to prove this, but we will follow Pólya's line of reasoning to show that $V - E + F = 2$ and $\Delta = 4\pi$ are equivalent statements. In the course of doing so, we obtain a result that takes us far beyond the domain of convex polyhedra as we defined them in Chapter 4. In fact, what we will prove is that

$$\Delta = 2\pi(V - E + F)$$

and you should immediately see that this identity establishes the equivalence of $V - E + F = 2$ and $\Delta = 4\pi$. However, we will make no use of convexity in our argument, and we will not need to assume that our polyhedron is deformable into the surface of a sphere; it will suffice that it is constructed out of polygonal faces, where two faces are put together by "gluing" a side of one to a side of the other to form an edge of the resulting surface. This last condition has the following important consequence. Let S be the total number of sides of the faces of our surface; then

$$S = 2E$$

For example, a tetrahedron consists of 4 triangles and each triangle has 3 sides; thus $S = 12$, while $E = 6$. Or, for the dodecahedron, we have 12 pentagons and each pentagon has 5 sides; thus $S = 60$, while $E = 30$.

We are now ready to prove $\Delta = 2\pi(V - E + F)$. What we do is to count the *sum of all the face angles* (which we will call A) in two different ways. We first count by vertices. Now, since the angular deficiency at the nth vertex is δ_n, the sum of the face angles at the nth vertex is $2\pi - \delta_n$. Thus

$$A = \sum_{n=1}^{V} (2\pi - \delta_n) = 2\pi V - \Delta$$

We next count by faces. Now, if a face has m sides, then the sum of the interior angles is $(m-2)\pi$, since, as we showed in Section 2.1, the sum of the *exterior* angles is 2π. Thus, if our polyhedron has F_m m-gons among its faces, those m-gons contribute $(m-2)F_m\pi$ to the sum of all the face angles. We thus arrive at the key formula

$$A = F_3\pi + 2F_4\pi + 3F_5\pi + \cdots = \sum_m (m-2)F_m\pi = \left(\sum_m mF_m - 2\sum_m F_m\right)\pi$$

Now

$$F = F_3 + F_4 + F_5 + \cdots = \sum_m F_m$$

Also, each m-gon has m sides, so that the contribution to the number of sides from the m-gons is mF_m. Thus

$$S = 3F_3 + 4F_4 + 5F_5 + \cdots = \sum_m mF_m$$

We put together these last three formulas, along with the fundamental relationship $S = 2E$, to infer that

$$A = (S - 2F)\pi = (2E - 2F)\pi$$

Comparing this with the earlier formula for A, obtained by counting by vertices, we conclude that

$$2\pi V - \Delta = 2\pi(E - F)$$

or

$$\Delta = 2\pi(V - E + F)$$

as claimed.

We repeat that this result is very general and takes us well beyond the very restricted class of (convex) polyhedra that has been our main concern in this book. Of course, we have to allow "negative angular deficiencies" if we no longer insist that our polyhedra be convex.

An interesting nonconvex example is furnished by the 12-celled collapsoids of Sections 11.3 and 11.4. You may count the constituent parts to see for yourself that $V = 26$, $E = 72$, and $F = 48$, so that $V - E + F = 2$. The different kinds of angular deficiencies are listed next:

> 12 vertices, surrounded by 4 equilateral triangles, contribute an angular deficiency of $\frac{2\pi}{3}$ each,
>
> 8 vertices, surrounded by 6 equilateral triangles, contribute an angular deficiency of 0 each,

and   6 vertices, surrounded by 8 equilateral triangles, contribute an angular deficiency of $-\frac{2\pi}{3}$ each (that is, an angular excess of $\frac{2\pi}{3}$).

Thus we see that the sum of all the angular deficiencies for the 12-celled collapsoid is

$$12(\tfrac{2\pi}{3}) + 8(0) + 6(-\tfrac{2\pi}{3}) = 6(\tfrac{2\pi}{3}) = 4\pi$$

so that $\Delta = 2\pi(V - E + F)$ holds in this nonconvex case, too, as promised.

Another interesting example is furnished by our rotating ring of tetrahedra (see Section 10.1). We imagine each of the "linking edges" between adjacent tetrahedra, which enable us to rotate the ring, pulled apart into two edges; this is necessary in order to retain the relationship $S = 2E$. If our ring is made up of k regular tetrahedra, then you may verify that

$$V = 2k \qquad E = 6k \qquad F = 4k$$

so that $V - E + F = 0$. On the other hand, 6 equilateral triangles come together at every vertex, so that the angular deficiency at each vertex is 0! Again, we see that for the rotating ring of tetrahedra the formula $\Delta = 2\pi(V - E + F)$ is splendidly vindicated, this time in a case in which the surface is definitely *not* deformable into a sphere, since our ring has a different Euler characteristic, namely, 0.

In fact, the configuration we get when we pull apart the linking edges is what we would call a *rectilinear model of a torus* (or *bicycle tire*). That is to say, just as our Platonic Solids, if made out of a malleable material, could be deformed into the shape of a sphere, so could our rotating ring be deformed into a torus, usually depicted like this:

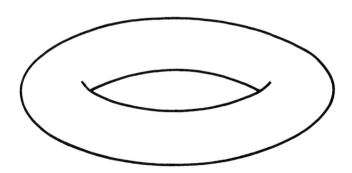

We come now to an interesting point about the relationship $\Delta = 2\pi(V - E + F)$, which will, we hope, appeal to the more mathematically (or more philosophically) minded of our readers. It is not just a matter of the two sides of our relationship, $\Delta = 2\pi(V - E + F)$, being equal. The Euler characteristic is obviously a *combinatorial invariant*, that is, it depends on the way our configuration is broken up into faces, edges, and vertices. In fact, we know that it depends on far less—for example, the value is always 2, provided only that the configuration can be deformed into a sphere. It is an example of a *topological* invariant (see the references at the end of this chapter). It is, at any rate, obvious that the quantity $V - E + F$ is unaltered if we distort the polyhedron somewhat (for example, if we massage a cube so that the faces are simply quadrilaterals). It is, however, by no means obvious that Δ is unaffected by such a distortion, since the angular deficiency at any particular vertex would certainly be expected to undergo change. Thus, $\Delta = 2\pi(V - E + F)$ tells us that, although Δ is defined by means of certain angular measures, it is in reality independent of those measures and depends only on the topological type of the polyhedron.

12.3   Some Combinatorial Properties of Polyhedra

We continue here to think of a polyhedron in the more general sense considered in the previous section; such a polyhedron may be described as a *closed, rectilinear surface*. Thus, as before, our surface has V vertices, E edges, F faces, S sides, and F_m m-gonal faces, so that

$$F = \sum_m F_m \qquad S = \sum_m mF_m \qquad S = 2E$$

We now prove two further basic relationships. They are so important that we call them *theorems*.

Theorem 1 $2E \geqslant 3F$; *equality holds if and only if each face is a triangle.*

Theorem 2 $2E \geqslant 3V$; *equality holds if and only if exactly 3 faces come together at each vertex.*

Before proving these theorems, we invite you to verify them for the Platonic Solids, using the table provided at the start of Section 12.2. You can also verify them for the convex deltahedra discussed in Chapter 4 and for the collapsoids discussed in Chapter 11.

Proving Theorem 1 is very easy in view of the relationships we gave just before its statement. Remember that, in forming the sums $\sum_m F_m$ and $\sum_m mF_m$, the number m takes values 3, 4, 5, ... Thus

$$\sum_m mF_m \geqslant \sum_m 3F_m, \text{ equality holding if and only if } F_4 = F_5 = \cdots = 0.$$

It follows that $S \geqslant 3F$, equality holding if and only if each face is a triangle. Since $S = 2E$, Theorem 1 is proved.

Our argument suggests that, to prove Theorem 2, we want to break up the vertices in a way analogous to that in which we classified the faces into m-gons for various m. Thus we write V_m for the number of vertices at which m faces come together, and we then want to prove that

$$V = \sum_m V_m \qquad 2E = \sum_m mV_m$$

Given these equalities, Theorem 2 is proved just as we proved Theorem 1. Since the first of these equalities is easily seen to be true, we concentrate on the second. We first remark that if m *faces* come together at a vertex, then m *edges* come together at that vertex (indeed, the analogy with the earlier classification of faces would perhaps have been better illustrated by talking of the number of edges coming together at a vertex rather than the number of faces). Thus if we count by vertices, we count in all $\sum_m mV_m$ edges, but each edge is counted twice since an edge joins two vertices. Thus

$$\sum_m mV_m = 2E$$

as claimed. If we want an analog of the idea of a side as used in Theorem 1, it is that of a *ray*, emanating from a given vertex. If R is the number of rays, then

$$\sum_m mV_m = R = 2E$$

But, although $R = S$, there is no sense in trying to think of each ray as a side or each side as a ray.

The proof of Theorem 2 is now easily completed (compare the proof of Theorem 1). We have

$$2E = \sum_m mV_m \geqslant 3\sum_m V_m = 3V$$

and equality holds if and only if $V_4 = V_5 = \ldots = 0$, that is, if and only if exactly 3 faces come together at every vertex.

In comparing the proofs of these two theorems, we find ourselves on the threshold of an exciting idea, that of a polyhedron and its *dual*. This is a pairing of polyhedra such that, if P and Q are dual polyhedra, then

$$V_m(P) = F_m(Q) \qquad F_m(P) = V_m(Q) \qquad E(P) = E(Q)$$

For example, the cube and the octahedron are dual; so, too, are the dodecahedron and icosahedron. The tetrahedron is dual to—itself! Actually, the duality is richer than we have indicated, but we have probably said enough!

The duality of the cube and the octahedron is a special case of the duality between a dipyramid having an n-gon for its equator and the prism having n-gons for bases. Likewise, the self-duality of the tetrahedron is a special case of the fact that every pyramid having an n-gon for a base is self-dual. You may check these statements in the following tables. (We haven't done all the work for you; you should work out the values of F_m and V_m, for various m, yourself.)

Having made these observations you might like to try to find other families of dual polyhedra. (Many exist.) After you have contemplated this question it may become clear why the duality between the dodecahedron and the icosahedron is seen as such a special relationship—there is no generalization available!

PAIRS OF DUAL POLYHEDRA

DIPYRAMIDS						PRISMS				
n		V	E	F		n		V	E	F
3		5	9	6		3		6	9	5
4		6	12	8		4		8	12	6
5		7	15	10		5		10	15	7
.		.	.	.		.		.	.	.
.		.	.	.		.		.	.	.
n		$n+2$	$3n$	$2n$		n		$2n$	$3n$	$n+2$

SELF-DUAL POLYHEDRA

PYRAMIDS				
n		V	E	F
3		4	6	4
4		5	8	5
5		6	10	6
:		:	:	:
n		$n+1$	$2n$	$n+1$

Let us close this section by drawing your attention to certain very concrete consequences of our theorems; these might be called *corollaries*.

Corollary 1 *If all faces on a surface are triangles, then the number of faces is even and the number of edges is divisible by 3.*

Corollary 2 *If 3 faces of a surface come together at each vertex, then the number of vertices is even and the number of edges is divisible by 3.*

Corollary 3 *A polyhedron (in the strict sense) cannot have 7 edges.*

We will be content to prove the third corollary, confidently leaving the proofs of the other two corollaries to you (remember that they are consequences of Theorems 1 and 2).

To prove Corollary 3, we suppose that $E = 7$ and hope that this will lead to a contradiction. Since $2E \geq 3F$, we have $3F \leq 14$, so that $F \leq 4$; similarly (using Theorem 2 instead of Theorem 1), $V \leq 4$. But then

$$V - E + F \leq 4 - 7 + 4 = 1$$

contradicting Euler's Formula, $V - E + F = 2$, for a polyhedron. Corollary 3 is proved.

Suppose you try the same argument with $E = 10$; we again assume we have a polyhedron in the original, strict sense. We have $3F \leq 20$, so that $F \leq 6$; likewise $V \leq 6$. Then

$$V - E + F \leq 6 - 10 + 6 \leq 2$$

Since, in fact, $V - E + F = 2$, we must have $F = 6$ and $V = 6$. From the equations

$$3F_3 + 4F_4 + 5F_5 + 6F_6 + \cdots = 20$$

$$F_3 + F_4, + F_5 + F_6 + \cdots = 6$$

we infer (subtracting 3 times the second equation from the first) that

$$F_4 + 2F_5 + 3F_6 + \cdots = 2$$

Thus $F_6 = F_7 = \cdots = 0$ and we have just two possibilities:

$$F_4 = 0, \qquad F_5 = 1, \qquad \text{giving } F_3 = 5$$

or

$$F_4 = 2, \qquad F_5 = 0, \qquad \text{giving } F_3 = 4.$$

The former possibility is realized by a pentagonal pyramid and the latter by the polyhedron shown on the right.

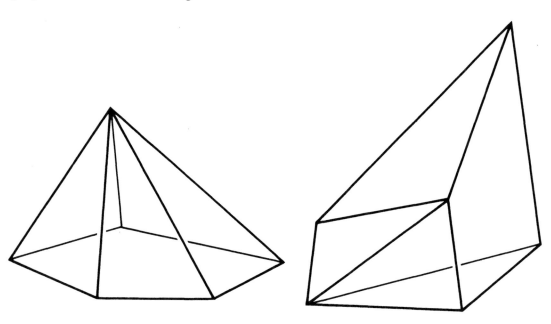

Are these polyhedra self-dual?

12.4 The Symmetries of the Cube

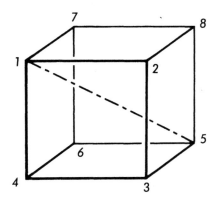

Diagonal 15 is D_1
26 is D_2
37 is D_3
48 is D_4

	Geometric description of the type of symmetry (where the numbers 1, 2, ..., 8 refer to vertices as labeled above).	Type of permutation of the vertices	Type of permutation of the main diagonals
Type 1	A 90° rotation in either direction about the axis through the centers of opposite faces; for example, through the centers of the front and back faces in one direction.	Two 4-cycles. In our example, (1234) (5678)*	One 4-cycle. In our example, $(D_1D_2D_3D_4)$

(There are 6 of these; find the other 5.)

Type 2	A 180° rotation through the centers of opposite faces; for example, through the centers of the front and back faces in one direction.	Four 2-cycles. In our example, (13) (24) (57) (68)	Two 2-cycles. In our example, (D_1D_3) (D_2D_4)

(There are 3 of these: find the other 2.)

*This would normally be written (1234) (5678); the vertical format is merely for typographical reasons.

Type 3

A 120° rotation in either direction about the axis through opposite vertices; for example, through vertices 1 and 5 in one direction.	Two 3-cycles and two 1-cycles. In our example, (247) (368) (1) (5)	One 3-cycle and one 1-cycle. In our example, $(D_2D_4D_3)$ (D_1)

(There are 8 of these; find the other 7.)

Type 4

A 180° rotation in either direction about the axis through the centers of opposite edges; for example, through the centers of edges 12 and 65.	Four 2-cycles. In our example, (12) (73) (84) (56)	One 2-cycle and two 1-cycles. In our example, (D_1D_2) (D_3) (D_4)

(There are 6 of these; find the other 5.)

Type 5

No movement. (This is called the *identity* or *zero rotation*.)	Eight 1-cycles.	Four 1-cycles.

(Of course, there is just one of this type.)

REFERENCES

Courant, R., and H. Robbins. *What Is Mathematics?* New York: Oxford University Press, 1978.

Hilton, Peter, and Jean Pedersen. "Discovering, Modifying and Solving Problems: A Case Study from the Contemplation of Polyhedra." *Teaching and Learning: A Problem-solving Focus*, ed. Frances R. Curcio. Reston, VA: NCTM, 1987, pp. 47–71.

What Next?

Let us whet your appetite for further studies by showing you another, remarkable way in which paper-folding leads to interesting number theory.

On page 56 we described, with the example of the regular 11-gon, how to create the folding instructions for a regular star $\{\frac{b}{a}\}$-gon. Here is a symbol that encodes the moves tabulated at the beginning of page 56.

$$11 \quad \left| \begin{array}{ccc} 1 & 5 & 3 \\ 1 & 1 & 3 \end{array} \right| \qquad (\bigstar)$$

Thus, we subtracted the fraction $\frac{1}{11}$ from 1 to get $\frac{10}{11}$, and bisected (halved) once to get $\frac{5}{11}$, and so on (see below). The $(\bigstar)$ is a particular example of a symbol

$$b \quad \left| \begin{array}{cccccc} a_1 & a_2 & \cdot & \cdot & \cdot & a_r \\ k_1 & k_2 & \cdot & \cdot & \cdot & k_r \end{array} \right| \qquad (\bigstar\bigstar)$$

where b is odd, each a_i is odd, relatively prime to b, and less than $\frac{b}{2}$; and

$$b - a_i = 2^{k_i} a_{i+1}. \qquad (\bigstar\bigstar\bigstar)$$

Moreover, there are no repeats of the a_i's (so that $a_{r+1} = a_1$).

We call r the *period* (of the paper-folding instructions) and, for convenience in the number theory we are about to explain, we set

$$k = k_1 + k_2 + \cdot \cdot \cdot + k_r.$$

You may now wish to turn again to page 56 to see if you understand how the symbol relates to the procedure that was used in order to obtain the folding instructions $D^3U^1D^1U^3D^1U^1$. If you can see this, you will probably be well on your way to understanding how to construct the general symbol. We'll explain the symbol $(\bigstar)$ fully now, just in case!

We will describe the process of constructing the symbol $(\bigstar)$ without actually referring to the tape. Start with $b = 11$ and $a_1 = 1$ (this will, in fact, uniquely determine the completed symbol) and write

$$11 \quad \left| \begin{array}{c} 1 \end{array} \right.$$

Now we compute: $11 - 1 = 10$, $\frac{10}{2} = 5$ (and STOP, because 5 is odd) and observe that this tells us that, in this instance, $(\bigstar\bigstar\bigstar)$ takes the form $11 - 1 = 2^1 5$, so record $k_1 (= 1)$ and $a_2 (= 5)$ to get

$$11 \quad \left| \begin{array}{cc} 1 & 5 \\ 1 & \end{array} \right.$$

Again we compute: $11 - 5 = 6$, $\frac{6}{2} = 3$ (and STOP, because 3 is odd), so that, in this instance, $(\bigstar\bigstar\bigstar)$ takes the form $11 - 5 = 2^1 3$, so we record $k_2(= 1)$ and

a_3 ($= 3$) to get

11	1	5	3
	1	1	

Repeating the process, we compute $11 - 3 = 8$, $\frac{8}{2} = 4$, $\frac{4}{2} = 2$, $\frac{2}{2} = 1$ (and STOP, because 1 is odd), so that, in this instance ($\bigstar\bigstar\bigstar$) takes the form $11 - 3 = 2^3 1$, so we record k_3($= 3$) and, since a_4 ($= 1$) would be the same as a_1, we STOP and draw the last vertical line to indicate that the symbol—which now appears as ($\bigstar$)—is complete. The numbers in the bottom row, when attached as superscripts to the sequence *DUDUDU* . . ., tell precisely how to fold tape, which can be used to construct the regular 11-gon (and, in fact, the regular $\{\frac{11}{2}\}$- and $\{\frac{11}{4}\}$-gons). Furthermore, we can see, knowing where the a_i in the top row came from on the tape, that this tape can also be used to fold regular star $\{\frac{11}{3}\}$- and $\{\frac{11}{5}\}$-gons.

Now for the surprise. The information in this symbol tells us the smallest number k such that either $2^k + 1$ or $2^k - 1$ will be exactly divisible by 11. In fact, in our particular example, $k = 5$ and the symbol tells us, since $r = 3$, that $2^5 - (-1)^3$, that is, $2^5 + 1$, is exactly divisible by 11—and that for no power of 2 less than the fifth can this be true, with $+1$ or -1. This is an example of the more general fact that for the variables in the symbol ($\bigstar\bigstar$), generated as described, for given b and any suitable a_1, it is always the case that

$2^k - (-1)^r$ is exactly divisible by b

and there is no smaller power l of 2 such that $2^l + 1$ or $2^l - 1$ is divisible by b. We call k the *Quasi-order of b mod2* and refer to the result as the *Quasi-order Theorem* (see the reference below for a proof of this theorem).

Here is a particularly interesting example of the Quasi-order Theorem—we'll explain why. Choose $b = 641$, and $a_1 = 1$ and construct the symbol. Try constructing the symbol for yourself before you look carefully at it, to give you some practice with the algorithm involving repeated use of ($\bigstar\bigstar\bigstar$)

641	1	5	159	241	25	77	141	125	129
	7	2	1	4	3	2	2	2	9

We can now calculate that $k = 7 + 2 + 1 + 4 + 3 + 2 + 2 + 2 + 9 = 32$, and observe that $r = 9$, so that the Quasi-order Theorem tells us that

$2^{32} - (-1)^9 = 2^{32} + 1$ is exactly divisible by 641!

We have just *proved* that the fifth Fermat number $2^{2^5} + 1$ is **NOT** prime. This fact was originally discovered by Leonhard Euler (see the footnote on page 7).

If you feel you are now ready for a proof of the Quasi-order Theorem and for further ideas in the same direction, along with some interesting questions that you could think about, you should consult

Hilton, Peter, and Jean Pedersen, "Geometry in Practice and Numbers in Theory," *Monographs in Undergraduate Mathematics 16* (1987), 37 pp. (available from the Department of Mathematics, Guilford College, Greensboro, NC, 27410).

Index